TWO GUYS FROM BARNUM, IOWA AND HOW THEY HELPED SAVE BASKETBALL

A History OF U.S. Patent 4,534,556

For A Break-Away Basketball Goal

Paul D. Estlund & Kenneth F. Estlund,

Inventors

FRANCIS B. FRANCOIS

Francois Press

Costa Mesa, CA

FRANCOIS PRESS

260 Victoria Street, Unit E4, Costa Mesa, CA 92627

Visit us on the web at http://www.francois.org

Manufactured in the Unites States of America

1 3 5 7 9 10 8 6 4 2

ISBN: 978-0-6151-8342-8

DEDICATION

To my late wife, Eileen, and our five children, all of whom felt the burden of the hours I spent at home and traveling working with the Estlunds to obtain and license their patent. A special thank you to my son Michael, for handling all of the details in getting this book published.

CHAPTER 1

BASKETBALL HAD A PROBLEM

This is the true story of the making and marketing of an invention by two American brothers who grew up on a farm in Iowa, about two miles from where I grew up. They are credited by many with helping to save the great game of basketball in the last decades of the 20th century, by making an invention and working to get it into widespread use.

The History of Basketball

The game of basketball was invented in America, by Dr. James Naismith. In 1891 he was given 14 days by his supervisor at the Young Men's Christian Association (YMCA) in Springfield, MA to create an indoor game that would appeal to active young men when the outside weather was poor. Naismith came up with the idea of mounting two wooden fruit baskets 10 feet high at each end of an open court, and using a soccer ball to play a game in which two opposing teams tried to score by putting the ball in the basket. Initially the baskets had bottoms in them, and the game had to stop after a goal to remove the ball from the basket. He later realized it was better to take away the bottoms so that the ball could fall through the basket and playing would be more continuous.

The game, which upon the proposal by one of the players was named "basket ball," caught on and spread to many other locations. The next year some women's colleges adopted basketball, and over the first years many universities and even the U.S. Navy adopted basketball, the U.S. Navy because it could be played onboard ships. As many universities took up the game, it became a major sport among the conferences of universities.

From the early days people would come to watch the games as an audience standing or sitting along the court's sidelines. During the first years no men were allowed to watch women playing basketball, however. In the early years basketball courts were enclosed in a cage made of netting to avoid the ball going into the spectators, and more importantly to discourage observers from trying to interfere with the game. This led to the use of the word "cagers" to identify the players. The use of cages was abandoned in 1929.

As the years went by many changes occurred in game equipment, rules of play and game strategy. One of the first changes was to adopt a wooden backboard upon which the basket was mounted, to prevent those in the gallery from touching the ball and to add rebounding and other elements of play; today backboards are commonly made of glass in professional and college games. In time woven nets mounted on circular steel rims replaced the wood baskets. The soccer ball was not considered ideal for the game, and innovators went to work to invent a better ball, made of leather, rubber or other materials and larger than a soccer ball. As result patents were obtained. One was U.S. patent 1,718,305 for a "Basket Ball," awarded to G.I. Pierce in 1929.

Basketball is now played by millions of people in America and around the world, and is watched by millions more. A lot of the play is informal, often outside. Primary and secondary schools commonly use basketball as a part of their physical education programs, and high school athletic associations establish the rules and playing conditions for high school games. And boys and girls clubs, YMCAs and other organizations across America offer basketball to their members.

Colleges play organized basketball under the National Collegiate Athletic Association (NCAA), leading to an annual national tournament held in March known informally as "March Madness." Finally, starting in 1898 professional basketball teams were organized in the United States and in other nations, and basketball is now an Olympic sport.

The object of the game is simple. The players are organized into two competing teams, and when a team has possession of the ball it works it down the court by dribbling and passing it from player to player. When a player with the ball feels he or she has a good shot, the player will shoot the ball toward the basket. If the shot is accurate it will go through the steel hoop and then through its attached net. The net slows the fall of the ball, and makes it easy to see that the ball actually went through. A skill pursued by players is to learn to bounce the ball off the backboard and through the hoop and finally the net.

The Late 20th Century Basketball Problem Solved by the Estlunds

Originally the basket mounted 10 feet high was well above the height of the players. But a profound change in the game occurred during the last decades of the 20th century when increasingly taller, heavier and physically stronger players started to jump into the air and then slam the basketball downwardly through the basket, sometimes deliberately grabbing and hanging on to the basket rim on their way down. These shots came to be called slam-dunks, and were initially popular with fans and increased attendance at games. But as time went on and more stronger, taller players adopted the slam-dunk shot, sometimes the result was a bent or broken basket or backboard, injury to a player, and/or long delays in the game because of the need to repair a damaged basket or backboard, or due to the need to physically assist an injured player. This was especially a problem in professional basketball, where for example Darryl Dawkins, as a member of the Philadelphia 76ers of the National Basketball Association (NBA), delighted fans by making these dunk shots, which sometimes resulted in a shattered glass backboard and bits of glass scattered on the playing surface.

The damaged and broken equipment, the delays to the game, and the possibility of injury to players caused basketball officials to ban the slam-dunk shot from basketball in the 1960s. When making its announcement in 1967 the NCAA rules committee spokesman said that it did so "as a result of 3,500 injuries and more than $250,000 in damaged equipment. Games and tournaments could not be completed and television contracts could not be fulfilled."

The fans and players were not happy with this, and attendance and interest especially in attending professional basketball games fell off. In 1976 the dunk shot was again made legal, and this started a new round of broken baskets and backboards, and game delays. The professional and college teams were faced with a problem to which there seemed to be no solution, and which was threatening the future of basketball.

High school players copied the popular dunk shots of the college and professional players, often with similar resulting damage to the equipment. This extended to pick-up games played outside, or in local gymnasiums or parks, where damage to baskets also began to occur. Clearly, the future of the game was facing a challenge, wherever it was played.

Figure 1 - A TYPICAL DESTROYED BASKETBALL GOAL IN AN OUTDOOR PARK, 1976

Enter the Estlunds

This was the problem that Paul D. and Kenneth F. Estlund (Paul and Ken) turned their attention to in 1976, and which they solved by inventing a break-away goal that keeps the basket level during normal play, allows it to bend down during a slam-dunk shot, returns it to playing position almost instantly, and which can be set to a specific detent release pressure so that the basket responds uniformly during the game, a feature that professional players told the Estlunds they wanted in such a goal. Goals constructed in accordance with their teachings have been adopted for professional and college basketball teams, and are now installed in gymnasiums and arenas across the United States and in other nations.

Estlunds Honored

In 1992 the importance of the Estlunds' invention was recognized by the Basketball Hall of Fame located in Springfield, Massachusetts, when during the 100th celebration of the invention of basketball it displayed the original working model of their break-away basketball goal. Ken had designed and constructed the model at his expense to further the obtaining of a patent, and to help popularize the invention. The Estlunds donated the original model to the Hall of Fame, where it remains.

Figure 2 - KEN AND PAUL ESTLUND (LEFT TO RIGHT) WITH THE FIRST MODEL OF THEIR BREAK-AWAY BASKETBALL GOAL, WHICH IS NOW IN THE BASKETBALL HALL OF FAME.

The Inventors and their Patent Attorney

The co-inventors are brothers, with Ken being older than Paul. They grew up on a farm near Barnum, Iowa, as did the author, who served as their patent attorney. All three of us helped our fathers farm, and were well familiar with machinery, springs and other mechanical devices. Ken and I went through grade school and high school together at Johnson Township Consolidated School in Barnum, Iowa. The building where we attended, built in the 1920s, is now gone, as is the presence of a high school in Barnum.

Paul attended the school in Barnum until the tenth grade, at which time he transferred to and was graduated from St. Edmond Catholic High School in Ft. Dodge, Iowa.

Ken and Paul were both involved in athletics during high school, including basketball, and they have followed the sport of basketball into their adult lives as fans of both the collegiate and professional games. They have each held employment that has allowed them to move about during the day by automobile, and which requires the knowledge and use of good business sense and tactics. In 1976 when the invention was made Ken marketed veterinarian pharmaceuticals in Colorado for Franklin Laboratories of American Home Products. Paul then did similar marketing in Illinois for Ft. Dodge Laboratories, and later moved to Colorado where he opened his own medical supply business.

Their Iowa farm background, their interest in and knowledge of basketball, their ability to move about and travel in their work, and their standing as successful businessmen all played a role in the making and exploitation of the invention.

The author is a registered patent attorney. At the time of the invention and up until November, 2004 I lived in

Bowie, Maryland. I hold an engineering degree from Iowa State University and a law degree from George Washington University and since 1961 have been registered to practice before the U.S. Patent and Trademark Office (PTO) in Washington, DC. As will be explained, I have served as patent counsel to the Estlunds since the invention was made in 1976. This history draws on selected documents contained in my voluminous files concerning the making of the invention, obtaining the patent, and marketing and licensing activity.

This patent was issued on August 13, 1985, and expired on August 13, 2002. A copy of the patent is included at the end of this book. Through the work of Ken, Paul and myself, and especially Ken, a number of licenses were negotiated that have generated substantial income to the three of us over the years. The Estlunds' invention has been a commercial success for us, for our licensees, and for the owners and players of the professional basketball teams.

Beyond commercial success, the Estlunds' invention has been important for the fans of basketball, especially fans of the professional and collegiate teams who now have all been using it since the 1980's. With a few exceptions each year the days of broken backboards and bent basketball goal rims is over, and fans can enjoy the game with its often-spectacular slam-dunk shots.

It is not an exaggeration to state that the Estlunds' invention helped save basketball, as a vital, enjoyable game.

For some time after making their invention the Estlunds, seeing no competitors, believed that they were the only inventors to tackle this basketball problem. But as will be discussed later the passage of time revealed that some other inventors also worked on the problem, and indeed new ideas are still being advanced. It was only the work of the Estlunds, however, that was finally adopted by the collegiate and professional basketball communities, after a trial period in actual games by the NBA in 1980.

Faced with a new round of broken baskets and backboards after the slam-dunk was reinstated in 1976, the NBA decided to test the break-away basket concept under game play. They notified those offering such goals of their plan, and many wanted to take part. The NBA picked three goals for testing under game conditions. One of those was the basket manufactured by Toss Back Inc., of Dorrance, Kansas, which was covered by the Estlunds' patent claims. This was the goal finally approved by the NBA. One of the remaining two baskets tested was also covered by the Estlunds' patent claims.

The importance and value of the Estlunds' invention to the basketball community would be enough reason for me to write this book. But I also have other reasons, including providing education and enjoyment to all who may wonder what inventors do, how patents are obtained, and steps that can be undertaken to receive royalties from an invention. There are chapters on the making of the invention, the obtaining of the patent, and the marketing efforts used to establish a royalty flow.

I handled many patent applications in my years of patent law practice, but this one holds the record for twists and turns and the time it took to obtain the patent. This is another reason for my writing, again for the enjoyment and education of those who have an interest in patents.

Finally, this book will help explain to our families those many hours expended by Ken, Paul and myself when we said we were "working on the patent and the invention." The three of us took many trips to meet each other, to confer with others, and to market the invention, in addition to many hours spent on the telephone. But it was all worth it.

CHAPTER 2

ABOUT U.S. PATENTS

Patents are at the heart of the American free enterprise system, and have been throughout our history. The U.S. Constitution established the right of inventors to obtain a patent of limited term to give them protection while they exploited their patented invention. During this term the patentee can prevent others from making, using or selling the patented invention. In exchange for this protection, the patentee agrees to transfer the knowledge contained in the patent to the general public after the term expires.

The term for patents has traditionally been 17 years, although in certain rare circumstances this can be extended. Thus, as of August 13, 2002, anyone can now make, use or sell the Estlunds' invention, and no royalty needs to be paid to them.

An inventor can follow any of several courses to exploit his or her patented invention. The invention can be sold outright to another party, for an agreed upon payment. The inventor can go into business to produce and sell the invention, rather than contracting with others or selling the invention. And the inventor can also license others to utilize some or all of the patent rights to the invention, in exchange for periodic royalties, a one-time royalty payment, or other consideration. The Estlunds considered all of these alternatives at one time or another, but in the end the bulk of the money earned by them from the invention came from licensing their patent to several manufacturing firms in business to build and market athletic equipment.

All patents begin with the preparation of a patent application, following rules established by the United States Patent and Trademark Office (Patent Office) under the authority given it by the Constitution and the U.S. Congress.

Patents can be granted to one person or, as here, co-inventors. The inventors hold title to the invention and the patent, unless they sell or license it.

In the Patent Office the application is assigned to a Patent Examiner, whose job it is to determine if the invention is useful, new and non-obvious, and whether or not anyone else has filed an application that appears to be for the same invention. If it appears that two or more inventors have filed on the same or very similar inventions, the Patent Office can start what is called an interference proceeding, in which all the facts about the inventions and their origins are developed and a patent is usually finally granted to one of the contending inventors or co-inventors. An interference proceeding can be very expensive and time consuming, but sometimes cannot be avoided. In the lengthy application process of our invention there came a time when we thought it possible we would be placed in an interference proceeding, but fortunately this did not happen.

The key issues in an interference proceeding are which inventor first conceived of the invention, and which one first reduced it to patent. The date of conception is when the idea was thought of. Reduction to practice can be either the date a patent application is filed, or the date when a working model is built and operated.

As stated, the Patent Office has established what a patent application must contain to be acceptable, including a written "specification" that describes the invention so that another person might practice it, and a drawing of the invention, if one is necessary to explain the invention and how it functions. The specification is followed by one or more "claims," which are like a deed to a piece of land in that they describe what the inventor believes is the invention. Finally, the inventor signs an "oath" or "declaration," in which he affirms that the invention is his and that it is accurately described.

The patent application is sent to the Patent Office, together with the fee established by the Congress. The Application Branch of the Patent Office assigns it to one of its Patent Examiners (the Examiner), who places it in his or her log for action. It can be one or two years before the Examiner first takes up the application, and when he or she does the first thing done is to review the application for any technical flaws.

Assuming the application is found to be in proper form, the Examiner then launches a search of Patent Office records to locate any relative prior art that might show that the invention described in the application is known, or would be obvious to one familiar with the field. Such prior art might be existing U.S. or foreign patents, or articles from magazines or books. The Patent Office maintains files of such materials, classified into different categories to facilitate the search by the Examiner or by outside persons, such as attorneys and ordinary citizens.

If the Examiner finds a patent or other document, called a "reference," that clearly shows the invention as described in the claim or claims is old, using that reference he will send the applicant an "Office Action," in which he will reject the affected claims over the reference, and set forth the reasons for the rejection. At the same time, if there are any technical problems that need attention, they will be set forth. If the Examiner cannot find any references and all is in order, he will normally "allow" the claim or claims, and forward the application to the issue branch.

Often what occurs is that no one reference is found that clearly shows the invention, and instead the Examiner locates several references that he believes if combined would in his judgment anticipate the claims. The argument is that it would be obvious to combine the references, to produce the invention described by the patent claims.

Once an Office Action has been received, the applicant normally has three months to respond to the Examiner's position. The response can simply be an argument that the Examiner is wrong, because he has misinterpreted the invention and/or the cited references, or for other reasons. At this stage the applicant's attorney may choose to file affidavit evidence supporting his arguments. The attorney may also amend the claims presented to the Patent Office, to avoid the teachings of the cited references. Once the applicant files a response, the application again goes on the Examiner's docket for action.

If the rejection is repeated and made Final, then an appeal can be filed with the Patent Office's Board of Patent Appeals (BOA), which can sustain or reverse the Examiner, in whole or in part. If the BOA fully sustains the Examiner's rejection, then it is possible to appeal to a U.S. Court.

All the proceedings in the Patent Office were confidential when the Estlunds' application was being examined, and the Patent Examiner could not identify or talk about pending applications made by others. The Examiner assigned to our application was Paul Shapiro, and he proved to be a determined man who put us through a very lengthy prosecution period, longer than any other I had experienced in some 40 years of practicing patent law.

This brief description of the U.S. patent system is of course not complete. For example, a fee was established to keep a patent in force and is now in effect on all new patents. We were not subjected to this fee, because our application was exempted under a grandfather provision of the law creating the fees. There are many other provisions of patent law and practice, but this should suffice to place the Estlunds' application in its setting.

CHAPTER 3

MAKING THE INVENTION

In late summer, 1976, Paul was living in Illinois. In driving around the Bloomington area he observed that a number of outdoor basketball goals in parks had been bent downward, apparently because of efforts by players to slam-dunk the basketball. He was disturbed by the damage. In early September during a telephone conversation with Ken, who lived in Parker, Colorado, Paul reported on what he had seen, and said that something needs to be done. After some discussion, Ken remembered the break-away plows that are used on the farm. He said he told Paul we could put a hinge on the basketball goal, mounting it so that it could break-away when a player hangs on the goal, and then use a spring to restore it to playing position. Thus was born by the two brothers the concept of their invention.

There then followed many activities, including the following.

September, 1976

Ken and Paul conferred on their concept, and agreed to look into patent protection for it. Ken remembered that I was a patent and trademark attorney, and contacted me by telephone to discuss the invention and to ask how to proceed. I recommended that as a first step a preliminary patentability search should be conducted in the U.S. Patent and Trademark Office, and described what is entailed in preparing, filing and prosecuting a patent application, including approximate costs.

Ken also said that he would like to get a trademark on Slam-Dunk, and we talked about the procedures for doing so. After reviewing the patent process I recommended that Ken and Paul write to me, enclosing copies of any sketches they had made to document their invention, and any written materials. This resulted in a letter written to me by Ken on September 25, 1976 to confirm our conversation, and including a sketch of the idea, a copy of a similar re-set spring system used on a farm plow, and a list of 7 points reflecting the advantages of our proposed safety basketball goal. He had the sketch prepared by a draftsman friend of his, Milt Haynes. The 7-point list was dated September 10, 1976.

October, 1976

On October 4, 1976, Paul mailed a letter to me with two sketches, one dated August, 1976 and the second September 27, 1976. The sketches showed two different ideas for the "Safe-T-Dunk" project we had discussed. Paul's September 27 sketch called for a spring loaded trip device that would activate upon application of a preset weight, and the sketch carried the comment that the breakaway basket must give a true bounce. The same sketch cited four benefits of the invention.

In a telephone conversation with the inventors, they requested that I proceed with a preliminary patentability search, and the issue of costs was discussed. Ken and Paul said they would be willing to give me a percent share in the invention for my services, and I said I would think about it. Over the years I had avoided becoming a partner with an inventor. But in this instance, I viewed Ken and Paul as long time family friends, and decided I would do it. I so advised them, and then proceeded to make a preliminary patentability search in the Patent and Trademark Office.

On October 5, 1976 I wrote to the Estlunds on the results of my search on the "Safe-T-Dunk Basketball Basket", enclosing copies of four patents that showed breakaway devices, but no breakaway basketball goal. I opined that based upon the search results, it should be possible to get worthwhile patent coverage on a break-away, re-settable basketball basket. I then went on to further discuss the obtaining of a patent, including the expected fees. I again agreed to consider working on the matter for a share of the invention, and said that if a patent application is to be prepared the next step was to concentrate on the preparation of adequate drawings. After further telephone conversations, Ken decided to come to Washington to meet with me, to discuss the principal and alternate concepts for the invention so that patent drawings could be prepared.

Ken came to Washington, and we reviewed different possible arrangements for the invention. After meeting with Ken I took the results of this meeting to James Lucht, my patent draftsman, and worked with him to produce a set of penciled drawings.

Figure 3 - FRANK FRANCOIS (left) AND KEN ESTLUND AT THEIR NOVEMBER, 1976 MEETING IN WASHINGTON.

Figure 4 – FRANK FRANCOIS AND KEN USTLUND IN 1945 GRADE SCHOOL PHOTOS TAKEN AT THE BARNUM SCHOOL, BARNUM, IOWA.

November, 1976

Ken wrote to me, forwarding a copy of the 1976 NCAA rules, and an advertisement by Illinois State stating, "The Dunk is Back!" Ken said that ads like this "sure will help the marketing of the product -- but I guess you'll agree that it also will require us to move as fast as possible to be first with the most."

December, 1976

On December 1, 1976, I wrote to Ken and Paul, forwarding prints of the penciled drawings for what was now called the "Breakaway Basketball Goal". I explained the drawings in detail, and asked for any comments from Paul and Ken. I also stated that I would begin preparation of the patent specification and claims, while waiting for comments on the penciled drawings

A few days later, Ken telephoned me and said he thought the drawings were fine. I asked for written comments, and Ken then wrote a letter on December 8 stating that "Paul and I are very pleased with the drawings." He then asked if I could provide a "rough copy" of the application by December 26, since Paul would be in Denver then for the holidays. Ken also asked for another meeting in Washington in late January, to discuss marketing of the invention. He also said that "a brief contract to bind our agreement would be in order," and that the shares would be 10 percent for me, and 45 percent each for Paul and Ken. In a post script to the letter, Ken said that "For filing, contracts, etc., put Paul's name first, then mine as co-inventors."

On December 28 I sent to Ken a draft in two copies of a patent application, and a draft agreement between the three of us. Ken responded stating that he and Paul agreed that the draft application was in order, as well as the three-party contract. I then proceeded to have the drawings inked and to place the application in final form.

January, 1977

On January 11, 1977, I mailed the completed application, contract and assignment to Paul for his execution, after which he was to send them to Ken, who would sign the documents and return them to me. The letter also included a statement dividing the drawing and filing fee cost among the three parties in accordance with the contract. The executed patent application was back in my hands by late January, and it was filed in the U.S. Patent and Trademark Office on January 27, 1977. The agreement required some further discussion, and was not signed until March 3, 1977.

Among the content of the application is an abstract of the invention as required by the Patent Office, which reads as follows:

ABSTRACT

"A basketball hoop is provided with supporting arm structure which is pivotally connected to a backboard, so that the hoop is movable downwardly from a normal, horizontal position to a break-away position. A detent mechanism is arranged to releasably secure the hoop in its normal horizontal position, but will disengage when a force of predetermined value is applied to the hoop. The hoop is then free to pivot downwardly into its break-away position. A reset mechanism is provided to return the hoop to its normal, horizontal position upon abatement of the applied force, and the detent mechanism is designed to be automatically re-engaging during such return movement."

CHAPTER 4

PROSECUTING THE PATENT APPLICATION

ROUND ONE

With the filing of the application, there began what turned out to be an unusually long prosecution period before the grant of a patent, the longest I ever experienced in over 40 years of practice. Over eight years lapsed between the filing of the application and the issuance of the patent on August 13, 1985. The prosecution of the patent application was difficult, and required two appeals to the Board of Patent Appeals in the U.S. Patent and Trademark Office.

A key reason for the long period of prosecution was that the Examiner, Paul Shapiro, had several other patent applications disclosing break-away basketball goals pending before him, in addition to ours. Under the secrecy rules of the U.S. Patent and Trademark Office there was no way we could have known this in 1977. As the years went by, the existence of some of those other applications became known to us. Some of the key dates and events relating to proceedings in the application were as follows.

January, 1977

The application was placed on file containing 14 claims that identified what we believed was patentable.

A Filing Receipt was received from the U.S. Patent and Trademark Office, advising that the application was received, and that it was awarded a filing date of January 27, 1977, and Serial No. 763,221. I advised Ken and Paul of this by a letter dated March 18, 1977.

March, 1977

Ken Estlund sent me a report on a meeting he had held with President Paul Ahrend of the Miracle Co., a maker of sports and game equipment, during which they discussed the invention. Ahrend said they planned to build some models, which would be tested. This was one of the first of many such visits by Ken and on occasion Paul, to market the invention.

May, 1977

The Examiner issued a first Office Action on the application on May 6, citing 8 references, and rejecting all of our 14 claims. No one reference showed our invention, however, and the Examiner instead resorted to combining elements of 4 or more of his cited patents, stating that it would be obvious to so combine them. I reported the Office Action to Ken and Paul on May 18, 1977, noting that a reply was currently due in three months, or by August 8, 1977. I said that I did not agree with the Examiner, and that I proposed to argue against the rejections without amending the claims. I also said that I thought more than the three months allowed might be needed to prepare a response, and that if this was true I would obtain a one-month extension, which was done on August 8, 1977 I suggested to Ken that the building of a working model should be completed before replying to the Office Action, to be certain that the claims covered it, and for use with the Examiner. Ken proceeded with this, turning to his friend Guy Williams, President of Franklin Labs, who had the model constructed from the drawings. Paul and Ken were both asked by me to provide materials to submit to the Examiner showing the uniqueness and importance of the invention, and they both did so.

September, 1977

An Amendment was placed on file on September 3, 1977, in which 3 additional claims were added to the application, and the reasons why claims should be allowed were argued at length. This Amendment was accompanied by several exhibits, including one comprised of material from copies of "Sports Illustrated," the "Chicago Sun Times" and two issues of a newspaper from Bloomington, Illinois, all provided by either Paul or Ken, and cited to show the importance of the need to solve the slam-dunk problem.

A copy of the current catalog from Miracle Recreation Equipment Company sent by Paul showing the kind of baskets then available, and a copy of a photograph also provided by Paul showing a current basket damaged in a slam-dunk shot were also submitted, along with a declaration from Kenneth E. Hughes, a friend of Ken who was then the head basketball coach for one of the largest high schools in Denver and to whom Ken had shown the invention. In his declaration, prepared with Ken's help, Coach Hughes stated that from his experience, he was not aware of anything like the invention, and that he believed the invention was valuable in solving the slam-dunk problem.

Ken had also obtained a signed statement from Robert J. Travaglini, the trainer for the Denver Nuggets professional basketball team. He had been shown the invention by Ken, and discussed it with key members of the Nugget's staff. In his statement he testified as to the uniqueness and value of the invention.

Also submitted was a declaration I had obtained from Hugh B. Robey, then the Director of Parks and Recreation for Prince George's County, Maryland. He had long experience with basketball equipment and approved purchases for a large park system. Mr. Robey was cast as an expert witness on sports equipment. Like Coach Hughes, Mr. Robey stated he was unaware of anything then available like the invention, and that "in my opinion, the Estlunds' invention is one of great merit, designed to truly solve the problems associated with the dunk shot."

The filing of the Amendment was reported by me to Ken and Paul, in a letter dated September 3, 1977.

December, 1977

During 1977 in the course of their talking with others in efforts to market the invention, Ken and Paul began to hear of other similar baskets being on the market. Paul wrote to me on September 9, 1977, stating that he had talked with several sources, but could get no specific information on such a basket. In September, while I was in Seattle, WA I found an article and picture in the September 24, 1977 "Seattle Times" showing Coach Chuck Randall of Western Washington University with a model of his "Slam Dunk Rim." I sent a copy of this clipping to Ken and Paul on October 29, 1977.

In early December, 1977 I visited Ken at his home in Parker, CO and discussed the marketing situation and the emerging evidence of other baskets on the market. Ken gave the author a copy of a brochure he had obtained from Coach Ken Hughes, who in turn had gotten it in the mail from a company called Slam-Dunk Rim, Inc., in Bellingham, WA. Paul joined us while I was in Denver, and Ken and Paul related their failed efforts to contact Coach Randall. During this visit I also presented a set of drawings based on the model Ken had constructed, dated December 5, 1977, and intended for use in marketing the invention.

Figure 5 - KEN ESTLUND (left) AND FRANK FRANCOIS MEET IN KEN'S HOME IN PARKER, CO TO REVIEW THE BREAK-AWAY BASKET MODEL AND THE STATUS OF THE PATENT APPLICATION

There were also discussions with Ken and Paul in Denver about another possible design for the break-away basket, centered on the backboard, and I agreed to develop some sketches of this for a possible second patent application we had been discussing. I wrote to Ken on December 19, 1977, after the visit, to provide a report on my visit to the U.S. Patent and Trademark Office when I talked briefly with Examiner Shapiro, and searched for evidence of other related applications, none of which was found. With respect to the Examiner, he could of course tell me nothing in light of the secrecy rules he worked under. My visit laid groundwork for an interview with him on our application, if needed.

I also reported on a trademark search I had done on "SLAM DUNK", which turned up one reference that had nothing to do with basketball.

In my discussions with the inventors they indicated their desire to get a trademark on "SLAM-DUNK", for marketing their invention. I explained that marks must be used in trade to be registered, and noted that at this stage we had no usage to base a claim on. Ken did proceed to get stationary with "SLAM-DUNK" on it, and used it where he could. But in time events passed us by, and no action was taken to file an application to register the mark. The reference I had found would probably have been cited against us, if we had filed an application.

January, 1978

A second Office Action was issued by the Examiner on January 3, 1978, in which all of the claims were Finally Rejected, with the Examiner making lengthy counter arguments responding to those I had advanced. I reported the Office Action to Ken and Paul on January 7, 1978, in a lengthy letter that discussed possible courses of action, from responding to the Examiner through noting an appeal. In this letter I also discussed progress on the proposed second application, and reported that I had developed a modification, a drawing of which was enclosed.

Ken responded to the January 7 letter on January 15, 1978, approving the course of action I had presented in my January 7 letter, which included as a first step responding to the issues raised by the Examiner, using whatever new back-up materials we had, and if this failed proceeding with filing an appeal with the Board of Appeals. Ken also enclosed several items that I had requested for the response, including three photographs of the working model, one showing Ken, Paul and Denver Nuggets player Bobby Jones standing beside the basket, a second showing Jones having tripped the rim, and a third showing the rim restored to its playing position. He also forwarded the original of the Slam-Dunk brochure.

Figure 6- PROFESSIONAL BASKETBALL PLAYER BOBBY JONES, THEN OF THE DENVER NUGGETS, IS SHOWN HAVING TRIPPED THE ESTLUNDS' BASKET

April, 1978

I filed a Response to the Office Action, including the three photographs sent by Ken, and arguing against the rejections. In addition, the Slam-Dunk brochure was placed of record, and used in the argument for allowing the claims. I reported the amendment to Ken on April 3, 1978, with the request that he pass on a copy to Paul. Also enclosed were sketch drawings for the second application. On April 9, 1978, Ken wrote agreeing to noting an appeal, if necessary. He also approved the sketch drawings for the second application

The Examiner replied to my amendment on April 12, 1978, with no change in his position. This left an appeal as the next recourse. I reported this development to Ken, with the request that he tell Paul.

On April 28, 1978 I filed an appeal of the Final Rejection to the Board of Appeals, on all of the claims. This was reported to the inventors in a letter dated April 29, 1978, in which I also confirmed that I would be in Denver again in early May.

May, 1978

Paul, Ken and I met again in Denver on May 7, and discussed all aspects of the situation. I outlined the next steps, including my preparing a lengthy brief for the Board of Appeals. I said that I felt that the Examiner's references were poor, and didn't really relate to our breakaway basket.

At this time we felt the world might be ours, and returned our discussion to the filing of one or two additional patent applications. We looked at the drawings I had done, and Ken and Paul discussed other possible versions of the invention. In particular, Ken and Paul discussed an economy version of a break-away basket, and they signed a sketch I made of it during our meeting. They felt this might be a third new patent application.

June, 1978

On June 26, 1978 I completed and filed the Brief on Appeal and a request for an oral hearing, which I reported to Ken on June 26, 1978, with a request that he get a copy to Paul.

October, 1978

At this point in our pursuit of a patent, the whole game changed and went in a new direction.

The normal course of action would have been for Examiner Shapiro to prepare and forward an "Examiner's Answer" responding to my "Brief on Appeal," with both documents then being sent to the Board of Appeals with the patent file.

Rather than filing an Examiner's Answer, which as noted would have been normal, the Examiner issued a new Office Action on October 3, 1978 in which a newly issued patent to Tyner was cited. The Examiner vacated the Final Rejections of the claims he had previously made, and then effectively reinstated them. He went on to further reject the claims using the Tyner patent, in combination with three others.

For the first time in all our activity, a reference had been cited that related to basketball and the slam-dunk problem - the Tyner patent. Up to this point we had no knowledge of Tyner's work. His full name was Frederick C. Tyner, of Raleigh, NC, and on March 25, 1978 he was issued U.S. patent 4,111,420 for an "Energy Absorbing Basketball Goal/Backboard Unit."

The Tyner patent file showed that his application was filed on July 19, 1976, over six months before we filed, and that it was issued on September 5,1978 by Examiner Shapiro. It dealt with much the same problem as does the Estlunds' invention, and was clearly a good reference against us. However, Tyner did not include key elements of our invention, and accordingly I believed that we could still get a patent for our invention. I wrote about all this to Ken in a lengthy letter dated October 7, 1978, with a copy to Paul. In my letter I expressed my belief that it might still be possible to obtain some worthwhile narrow claims, but that the broad coverage originally hoped for was no longer possible. I offered a set of options, including developing revised claims, and the possible filing of new applications. The counsel of Ken and Paul was for me to move ahead and respond to the new rejections and the Tyner patent.

December, 1978

I prepared and filed a "Request for an Extension of Time" to respond to the October 3, 1978 Office Action, which was granted.

February, 1979

I completed and filed an Amendment on February 3, 1979 in which Claims 1-4 and 9-17 were canceled, and the remaining claims amended to overcome the Tyner patent while protecting key features of the Estlunds' invention. The approach was developed in conversations between me, Ken and Paul. The filing of the Amendment was reported in a letter to Ken dated March 5, 1979, along with a copy of the Request for Extension of Time filed in December. Ken was asked to provide copies to Paul

March, 1979

In the spring of 1979 we gained more knowledge of Coach Randall's basket, and judged that he had copied ideas from our invention. In response on March 16, 1979 I filed a "Supplemental Amendment" adding three more claims to the application, Claims 18-21, which clearly read on the Randall basket. On April 21, 1979 I wrote to Ken, with copies of the Supplemental Amendment. In the same letter I enclosed for Ken and Paul a draft letter to Coach Randall, suggesting we explore working together

April, 1979

The Examiner mailed another Office Action on April 23, 1979 in which he cited two additional supplemental references, and issued a Final Rejection on all the claims. I reported this to Ken in a letter in which I stated that I would wait to note the appeal until just before the deadline, to allow time for us to seek an agreement with Coach Randall.

With regard to Coach Randall, on May 4, 1979, Ken wrote to me advising that my draft letter was acceptable to both him and Paul, and he said it had been mailed the day before. The letter was sent on Slam Dunk, Inc. letterhead that Ken had prepared, and was signed by both Ken and Paul. As of May 24, no reply had been made to the Coach Randall letter, as I noted in my letter to Ken.

July, 1979

On July 23 I prepared and filed a Notice of Appeal from the April 23, 1979 Final Rejection, and reported it to the inventors.

September, 1979

I prepared and filed a Brief on Appeal in the application and a request for an oral hearing before the Board of Appeals. On the same date I wrote to Ken, sending the usual two copies of my Notice of Appeal and the Brief on Appeal, which had been reported to Ken by telephone

November, 1979

The Examiner filed his "Examiner's Answer" on November 17, 1979. It was unusual in that the Examiner entered a new ground of rejection for Claim 8, citing a new reference, a patent to Court. This action allowed for a response, and the Examiner set a two month time for this to be done.

The Court patent, like most of the other references cited by the Examiner, had no connection with basketball, contrary to the Examiner's argument. This was reported to the inventors.

January, 1980

I filed a Reply Brief on January 4, 1980, responding to the new rejection of the Examiner. On January 4, 1980, I reported the filing of the Reply Brief to Ken, with copies for him and Paul. In the letter I referred to another meeting I had with Ken in Denver, during which the position of the Examiner and my planned responses were discussed.

February, 1980

Ken sent me a letter on February 20 that enclosed a copy of a letter to him from Joe Axelson, Director of Operations for the NBA, regarding what their standard was for break-away rims. In that letter Axelson stated that they would be testing a rim "manufactured by Ken Mahoney of Dorrance, Kansas." This was the first time that Mahoneys' name had surfaced, and Ken and Paul decided to make contact with him. This they did, and the meeting led to cooperation between Ken Mahoney and us.

I earlier discussed the 1980 NBA test, which in fact included three break-away goals, including the Mahoney goal. It turned out that both the Mahoney goal and another one of the three were ultimately covered by the claims of our issued patent.

April, 1981

I appeared before the three member Board of Appeals on April 6, 1981, and after describing the purpose and features of the invention presented an oral argument against the rejections of the Examiner. On May 20, 1981, the Board of Appeals delivered a unanimous decision in our favor, reversing the rejections on all of the pending claims. I immediately reported this to Ken and Paul, and we rejoiced in what we thought was a victory after a tough battle.

We assumed that Examiner Shapiro would now allow the application and issue the patent. But this was not to be.

Examiner Shapiro thought the Board of Appeals was in error, and chose to fight on. Specifically the Examiner mailed us a notice suspending further action in our application awaiting the issuance of yet another pending application that he would then cite against us. And so we had to sit and wait for many months.

CHAPTER 5

PROSECUTING THE PATENT APPLICATION

ROUND TWO

By now it had become evident to us that Examiner Shapiro had perhaps several applications before him dealing with break-away basketball goals. He made a strategic decision at this time to delay further action on our application until he could issue a patent on another invention, which he felt was earlier than ours and that would be a good reference against our claims.

And so we went into a holding pattern at the U.S. Patent Office, until mid 1983. But by no means were these idle months.

Enter Our Competitors

Since making their invention the Estlunds worked to find ways to take it to market. Early on they considered the possibility of establishing a manufacturing plant and sales force, but the capital needed for this was large. Instead, they decided to find others who would license our patent(s), and pay us a royalty. To find possible licensees Ken and/or Paul traveled many miles, visited many manufacturers, wrote letters and telephoned, talked with coaches and players, and attended NCAA March Madness and other organization's events. Some of these activities have been referred to above.

I contributed to these activities where I could and forwarded to Paul and Ken information that I found, such as the newspaper accounts of Coach Randall's invention. As their attorney, I also contributed to the outreach effort by writing such things as the letter to Coach Randall and other marketing items, and by drawing on the knowledge of athletic equipment makers I acquired from holding public office.

More about licensing activity later. For now, I want to focus on one company and one man that Paul and Ken met in their marketing work. The company was Toss Back, Inc. of Dorrance, Kansas, and the man is Kenneth J. Mahoney, the president of the company, who was mentioned above. Ken Mahoney was a member of the Kansas State team that won the national championship in 1951 and remained active in basketball, knowing several of the college coaches on a first name basis. He held some patents, and his major product was a basketball training device featuring resilient screens mounted on a frame that the player could bounce the basketball on, the screens then "tossing back" the ball. Ken Mahoney was active on the basketball scene in colleges and at the NCAA level, and he had also developed a break-away basket that included some features of our invention. Mahoney was marketing his break-away basket, and in that effort found some other companies in the nation also interested in the product. The work of the Estlunds came to his attention, and he agreed to meet with Ken and Paul.

On June 19, 1980 Paul wrote me a letter describing his trip to Kansas to see Mahoney. He said that Ken Mahoney had talked with him about a competing company, Slam Dunk Inc. of Seattle, WA, headed by J. Simonseth. Mahoney told Paul that Simonseth "is very political and has a lot of contacts and is in his opinion the front runner" in break-away goals. Paul's letter closed with "I feel if we could team up with Mahoney and his people we may all be a lot better off."

In their discussions with Mahoney, Paul and Ken filled him in on our basket and the fact that a patent was pending. They also discussed me, and the fact that I was in Washington where the U.S. Patent Office is located. Mr. Mahoney was having trouble getting his patent, also, and it developed that he too was facing Paul Shapiro. He was also aware of some other companies selling baskets like ours, and filled Ken and Paul in on some of them.

We also discovered that Ken Mahoney had a deceased brother, Elmo, who was listed as a co-inventor on his patent application. We had to take this into account when we entered into any agreements with Ken Mahoney.

Ken Mahoney came to Washington to meet me and discuss his application, and said it would be nice if I could help him get his patent while I was working on the Estlunds' application. Ken and Paul thought about the situation, and we explored the pros and cons of joining with Mahoney, in a team effort to get our patents and market the invention. Paul, Ken and I discussed what a partnership might look like, and I then drafted the outline of an agreement. I also sent a July 18, 1980 letter to Phillip H. Rein, Ken Mahoney's patent attorney located in Wichita, KS. My letter outlined some of the issues and possible approaches to an arrangement with Toss Back, Inc. Ken and Paul met with Mahoney and discussed the terms of the agreement. The inventors then reported to me, and I drafted a formal Agreement and sent it to Mahoney and his patent attorney, on August 9, 1980. Ken Mahoney again met with me in Washington before I prepared the August 9 letter, and we again reviewed the status of matters. After some discussion and a number of amendments the Agreement was executed and became effective on October 18, 1980. The parties included Toss Back, Inc. and Ken Mahoney, Ken and Paul, and me.

Under the Agreement the Estlunds and Mahoney executed assignments of their pending applications to Toss Back, Inc. Royalties from Toss Back, Inc. were to begin accumulating as of October 18, 1980, and the parties agreed to exchange full information about their inventions to each other, along with any marketing information. Finally, I was to be available to Toss Back, Inc. and Phil Rein to advise on their patent applications. There followed a complete exchange of information, and we evaluated our options for moving ahead.

On October 20, 1980 a new contract between Paul, Ken and me was signed, to replace the 1977 agreement, and to accommodate the Toss Back, Inc. arrangement. Under this agreement the Estlunds agreed to make me an equal partner with them with regard to any royalties.

June, 1981

After receiving the notice of suspension on our application I conferred with Phil Rein about the usefulness of my interviewing Examiner Shapiro, and we agreed it might be of value. After some difficulty in reaching him I held a telephone interview with the examiner in mid June, and advised him of the relationship between the Mahoneys' and Estlunds' applications and the fact they are now in common ownership. I stated we were concerned about the time this has taken, and urged him to move ahead. He said he would soon take up "all the cases", which was important information to us. It became clear to me that the Examiner was considering setting up an interference proceeding, and was working on a common claim that would address all the applications. Fortunately, this never happened. I conveyed this information to Phil Rein, Ken Mahoney, and Paul and Ken on June 13, 1981.

Recognizing that an interference proceeding might be in our future, Phil Rein and I both advised our clients of the situation, and urged that we move now to collect any hard evidence we had as to date of invention, first use in public, and similar information. Ken Mahoney and the Estlunds both responded to this plea, but as it turned out the evidence was not needed since no interference was declared.

Over the next 20 months we continued to wait on Examiner Shapiro, who had imposed the same delay on the Mahoneys' application as he had on us. We recorded the Assignment of the Estlunds' application, so that our two applications were in common ownership at the Patent Office. Ken and Paul continued trips to visit possible licensees and to spread the word about the invention, while Ken Mahoney and Toss Back, Inc. moved ahead with the manufacture and sale of their break-away basket ball goal, and their several other products. The press was carrying more articles and sales literature for break-away goals, and it was clear that several makers were now out there.

June, 1983

On June 1, 1983 Examiner Shapiro finally ended the hold on our application. He issued another Office Action in which he "reopened" prosecution of our application, and cited a new U.S. patent to Arthur H. Ehrat of Lowder, IL, Patent No. 4.365,802 for a "Deformation-Preventing Swingable Mount for Basketball Goals." It was issued on December 28, 1982 on an application filed July 26, 1976, and was a good reference against us. The Examiner cited the same patent against the Mahoneys' application, where its dates also were valid as a reference.

The Ehrat patent was for the basketball art, but like Tyner it did not show our invention. The Examiner recognized this, and combined others of his cited patents to be used with Ehrat to reject our claims. After reviewing the situation, I brought Paul and Ken up to date and advised that I still thought we could prevail. I advised that we should proceed with filing a response, and they agreed.

Ehrat proceeded to exploit his patent by undertaking an active marketing effort to obtain licensees. One of those was Toss Back, Inc., after he had filed an infringement suit against it. Toss Back, Inc. agreed to pay a 3 % royalty, which upset Ken Mahoney.

Phil Rein and I exchanged views on the two rejections, and we concurred that we would both file responses in our applications. I proceeded to prepare an amendment, including the addition of some new claims and amendments to others. During the long hold by the Examiner Phil Rein and I took the time to review the coverage of all our claims in light of Coach Randall's invention and some other activity we were now aware of, to assure we protected our two inventions. On August 29, 1993 I filed my Amendment, which included extensive arguments against the Examiner's rejections. .

November, 1983

On November 18, 1983 the Examiner issued another Office Action in which he finally rejected all of the claims. In reviewing the Office Action I found that he had missed one claim. This caused me to file another amendment pointing out the Patent Office error. The Examiner responded on January 5, 1994 by correcting his error, and repeating the Final Rejection. All of this was reported to Ken and Paul, and Phil Rein. It seemed clear that we must again go to the Board of Appeals, and Ken, Paul and Phil Rein all agreed.

April, 1984

On April 4, 1984 I noted an appeal of the Final Rejection, and requested an oral hearing. I then proceeded to prepare a 22 page "Brief on Appeal," which argued against all of the rejections and the use of the Tyner and Ehrat patents. I conferred with Paul, Ken and Phil Rein on the brief, and filed it in the Patent Office,

August, 1984

On August 10, 1984 the Examiner mailed his "Examiner's Answer" responding to my "Brief on Appeal" In reviewing it I found the arguments I expected based on the cited references, and then noted that the Examiner had raised a whole new issue – implying that the Estlunds were not the true inventors. The Examiner did not name another party, but it appeared from his wording that he was referring to the Mahoneys' application, which was also before the Examiner. I reviewed this with Paul, Ken and Phil Rein, and we agreed that I should respond to the new argument.

To provide time to prepare and file the reply brief an extension of time was obtained on August 22, 1984. The reply brief was filed on September 24, 1984, together with a request for an oral hearing before the Board of Appeals. On December 18, 1984 the Board of Appeals responded, and set an oral hearing for February 11. 1985. The three-month period between September 24 and December 18 was unusual, but I found out that the reason was the Board wanted to assemble the same three members as decided our first appeal.

February, 1985

In preparing for the oral hearing, I conferred with Phil Rein and we decided that I would get into the Toss Back, Inc. application, to try and tie things together and to overcome the Examiner's last ditch effort to question inventorship.

On February 11, I appeared before the Board of Appeals and argued our case. Now it was back to a waiting game that ended on March 19, 1985 when the Board of Appeals rendered its opinion, completely overturning the Final Rejections of Examiner Shapiro. The question now was, what would Mr. Shapiro do this time when the Board overturned him?

May, 1985

On May 7, 1985 Examiner Shapiro bowed to the Board of Appeals, and issued a Notice of Allowance for all of our claims. VICTORY AT LAST!

All that now remained to do was payment of the Issue Fee on June 3, 1985. Doing this caused some problems with Ken Mahoney, who had not returned a form I had sent him to complete, establishing us as "small inventors", which carried a much smaller issue fee. Working with Phil Rein, we finally got this worked out. Our patent finally issued on August 13, 1985, as Patent No. 4,534,4556.

June, 1987

Ken Mahoney and Phil Rein continued to pursue their patent application through appeal of the final rejection, and once again Examiner Shapiro was overruled. The Mahoneys' patent was issued on June 30, 1987 to Kenneth J. Mahoney and his deceased brother and co-inventor Elmo J. Mahoney, as patent No. 4,676,503 for a "Break-Away Basketball Goal Apparatus." The Estlunds' patent claims clearly covered the Mahoneys' invention, which Toss Back, Inc. had been actively selling for a few years.

CHAPTER 6

GETTING THE BENEFITS OF THE PATENT

With our patent now issued, the time had come for the Estlunds to seek revenue from their invention. As discussed earlier, patents can be commercialized in essentially three ways; outright sale to someone, going into the business of manufacturing and selling the invention, or licensing others to make and sell the invention for royalties. Ken and Paul were willing to discuss outright sale, but no viable opportunity came up. Ken, using his contacts with the Denver Nuggets NBA franchise, developed a proposal to them, but the interest just wasn't there.

The last of these three options seemed to be the most logical way to proceed, especially since the Estlunds-Toss Back, Inc. agreement of 1980 included a provision that once our patent issued Toss Back, Inc. would proceed to pay to the Estlunds and Ken Mahoney the royalties earned on covered break-way baskets that had been sold after October 18, 1980,

December, 1984

On December 20, 1984 I received a letter from Phil Rein alerting us to a developing situation between us and Ken Mahoney, with Rein suggesting that conditions had changed a lot since 1980 and our agreement should be changed. Mahoney was now paying a 3 % royalty to Ehrat, and was worried about others in the market, such as Coach Randall's basket. I responded on January 2, 1985, saying we are willing to talk about it but that no action was really needed for now.

March, 1985

On March 20, 1985 I wrote to Phillip Rein advising him that the Board of Appeals had acted favorably on our application, and reminded him that once our patent issues royalties going back to October 18, 1980 would be due to the Estlunds at the rate of 4% of the price charged by Toss Back, Inc to its customers. This demand for payment was not well received by Ken Mahoney, who as noted above believed that much had happened since the 1980 agreement and it had not turned out as he expected. We clearly had a problem, and Paul, Ken and I went to work with Phil Rein to see what could be achieved, short of our simply filing a suit for performance against Toss Back, Inc.

August, 1985

On August 16 I wrote to Ken Mahoney, advising that the Estlunds' patent had issued on August 13, 1985, and demanding a payment of royalties as required by our October, 1980 agreement with Toss Back, Inc. I also enclosed some accounting sheets, for determining the royalty due us. No reply was ever received.

October, 1985

In an October 11, 1985 letter Phil Rein sent a detailed letter containing Ken Mahoney's concerns. He was concerned about the years that had past, during which others entered this market. He was in fact a licensee under the Ehrat patent. He had expected our 1980 agreement to give him full control of the marker, even though this could not have happened with so many players.

Ken Mahoney also thought we had not really helped market the invention, and he had some other gripes. Their proposal essentially was: (1) to buy us out for $10,000; (2) to offset the royalty from Toss Back, Inc. by the 3 % royalty they paid to Ehrat; and (3) some other clauses relating to marketing. I shared the Rein letter with Ken and Paul, and we discussed about how to handle it. Ken and Paul decided a meeting with Mahoney was in order.

To prepare for the meeting, Paul, Ken and I drafted a position paper, comparing some options. On the royalty issue, our estimate was that $40,000 or substantially more might be due and that we would seek a mutually agreed to settlement. We also decided that we wanted our patent back, and that in exchange we could cross-license the two patents allowing each of us to grant a license on the other party's patent.

Toss Back, Inc. had Gary Mahoney hold discussions with Ken Estlund, and several proposals were made by him. All failed because the royalty payment was simply too low. I then wrote again to Phil Rein withdrawing all of our proposals, and including a new contract to replace that of 1980. This went back and forth for months, but finally in June, 1986 we completed an agreement that: (1) paid the Estlunds for back royalties; (2) reassigned our patent to the Estlunds; (3) cross-licensed the two patents, so that either of us could license the other's break-away basket, (4) granted a royalty free, non-exclusive license to Toss Back, Inc. of the Estlunds' patent; (5) and cancelled the 1980 agreement. This was the first substantial income we earned from our patent effort. I proceeded with some follow-up actions, including recording the re-assignment of our patent in the Patent and Trademark Office.

After this final arrangement with Toss Back, Inc., Paul and Ken increased their efforts to interest other companies in licensing our invention, with considerable success.   One of the things they first considered was the possibility of forming a corporation to hold the patent and receive any royalties. This concept appeared to have some legal and tax advantages, but in the end was not pursued. Instead, it was decided that any license agreements would require the payment of one-third of any royalty directly to Paul, Ken and me, and the agreement between the three of us was amended to provide for this.

1990

During 1990 Paul and Ken, and especially Ken, was busy contacting, visiting and telephoning companies in the basketball goal business. The first license we executed for royalties was with Basketball Products International, located in Seattle, WA.

The way we agreed to proceed was for Ken to make the contact and sales presentation, and discuss overall terms. Paul, Ken and I discussed the terms we might offer and seek in contracts, and reached agreement on goals.   Working within that broad outline, Ken worked to get the best terms he could. He then reported to me, and it was my task to prepare a license agreement, which was usually sent to the company. On some occasions those we contacted got their attorneys involved, and I worked with them to develop an agreement both sides could accept.

I always assumed that the customer's attorneys would make a review of our Patent Office file, and I welcomed this. One big advantage we had was that our Patent Office tribulations now made our patent a strong one that would probably hold up in court, if it came to that. Also, the length of time it took to get the patent worked in our favor, because our 17 years started after the break-away goal was in use and accepted.

The license for Basketball Products International, Inc. was executed on March 27, 1990, and included only the Estlunds' patent. The company was sold in later years to American Athletic, Inc. an Iowa company, which honored the BPI license.

In 1990 licenses were also executed by:

Bison Recreational Products (later Bison, Inc.) of Lincoln, Nebraska, under both the Estlunds' and the Mahoneys' patent - May 25, 1990

Schutt Manufacturing Company, under both the Estlunds' and the Mahoneys' patents - June 8, 1990

1991

Sure Shot Sports Inc. (became bankrupt and its assets were acquired by Huffy Sports Basketball Systems on July 22, 1997), under both the Estlunds' and the Mahoneys' patents - April 1, 1991. Huffy assumed our license agreement and paid royalties until its termination.

1996

Hyland Engineering - under both the Estlunds' and the Mahoneys' patents - March 12, 1996. The company failed near the end of the patent, and went under owing us considerable money.

1998

Thriller Manufacturing Company - under both the Estlunds' and the Mahoneys' patents - April 23, 1998

1999

Porter Athletic Equipment Company – under both the Estlunds' and the Mahoneys' patents – October 4, 1999

The typical terms included an up-front payment, and a quarterly royalty report of the number of covered baskets that had been sold, at an agreed upon fee for each. We usually had a cost of living (COL) adjustment every year. Ken acted as manager of our patent licenses, notifying the licensees of the amount and date of the COLs and following up with them on collecting royalties and other matters.

A FEW FINAL WORDS

My working relationship with Paul and Ken was good over the years. Ken was always available, and I could ask him to do anything that we felt was needed. He was almost always available by telephone, and if not I got a call back as soon as possible. Paul was a little harder to reach, especially when he moved from Illinois to Colorado and began life with a new wife. But I could usually reach him in time for decisions.

Both Paul and Ken made many important contacts in the basketball field, talking with players, managers, players and prospective licensees, and providing me with the information I needed to work with the Patent Office, Toss Back, Inc. and others. I have told of some of their calls and notes to me in this text, but only as examples – there were many more.

This patent did not make us wealthy, but did provide some welcome dollars. For me, this effort drew on all my talents and had many milestones that I will always remember. It also gave me a chance to work with Ken and Paul, and to enjoy many hours with them.

As to Toss Back, Inc. the computer reports that Ken Mahoney was involved with a failed bank in Kansas in the mid 1980s and was forced by the FDIC to sell out, except for his building. Also according to computer reports his son, Tom, and his wife have opened Pro-Bound Sports on the same site and have a catalog much like Kens. Mr. Ehrat had success with his patent, and deposited his records in the Archives Center, National Museum of American History, the Smithsonian Institution in Washington, DC. He had a history of lawsuits to convince firms to license his patent, and brought many companies on board.

It is my hope that this book will help show the value and importance of the patent system in helping inventors succeed with their efforts. We tend to think great names like Thomas Edison and industrial giants like Henry Ford and Bill Gates when we think about patents. But there are many, many small inventors like Paul and Ken Estlund who every year learn to utilize the patent system to successfully produce something of value to our lives. Admittedly obtaining a patent can be difficult, as was true for the Estlunds. But as we proved, success is possible.

What advice do I offer inventors and their patent attorneys? The first thing is to know the facts about your invention – ask questions, offer suggestions, and investigate any alternatives the inventor(s) explored. Second, prepare a sound patent application, with the inventor's involvement. Third, recognize that the Patent Examiner has a difficult job and will be as fair as he or she can – they in fact are your friend, not an enemy. Finally, explore all your options before giving up – be persistent, as you see the facts.

Francis B. Francois
November 11, 2007

United States Patent [19]

Estlund et al.

[11] Patent Number: 4,534,556

[45] Date of Patent: Aug. 13, 1985

[54] **BREAK-AWAY BASKETBALL GOAL**

[76] Inventors: Paul D. Estlund, 4174 S. Kalispell, Aurora, Colo. 80013; Kenneth F. Estlund, 8983 Inspiration Dr., Parker, Colo. 80134

[21] Appl. No.: **763,221**

[22] Filed: **Jan. 27, 1977**

[51] Int. Cl.³ A63B 63/08
[52] U.S. Cl. .... 273/1.5 R
[58] Field of Search .... 273/1.5, 105; 172/261–266, 269; 248/475 A, 475 B

[56] **References Cited**

U.S. PATENT DOCUMENTS

790,410	5/1905	Warne	172/265
1,194,006	8/1916	Fry	172/265
2,506,443	5/1950	Court	172/265
2,935,144	5/1960	Graham	172/265
3,194,555	7/1965	Humphrey	273/1.5 R
3,402,773	9/1968	Jennings et al.	172/265
3,468,382	9/1969	Quanbeck	172/264
3,795,396	3/1974	Kropelnitski	273/1.5 A X
4,111,420	9/1978	Tyner	273/1.5 R
4,365,802	12/1982	Ehrat	273/1.5 R

FOREIGN PATENT DOCUMENTS

604945	9/1960	Canada	172/264
1479428	3/1967	France	248/475 B
613310	11/1948	United Kingdom	172/269

Primary Examiner—Paul E. Shapiro
Attorney, Agent, or Firm—Francis B. Francois

[57] **ABSTRACT**

A basketball hoop is provided with supporting arm structure which is pivotally connected to a backboard, so that the hoop is movable downwardly from a normal, horizontal position to a break-away position. A detent mechanism is arranged to releasably secure the hoop in its normal horizontal position, but will disengage when a force of predetermined value is applied to the hoop. The hoop is then free to pivot downwardly into its break-away position. A reset mechanism is provided to return the hoop to its normal, horizontal position upon abatement of the applied force, and the detent mechanism is designed to be automatically re-engaging during such return movement.

5 Claims, 13 Drawing Figures

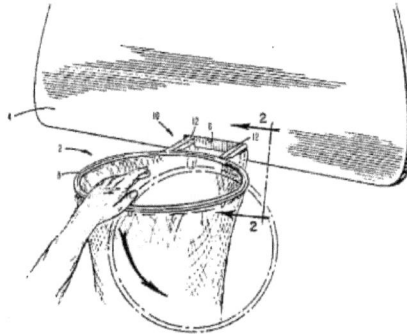

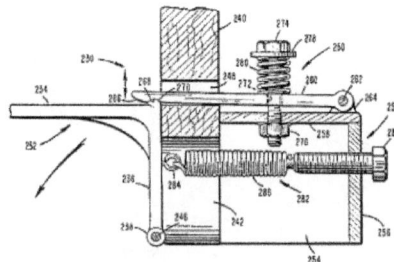

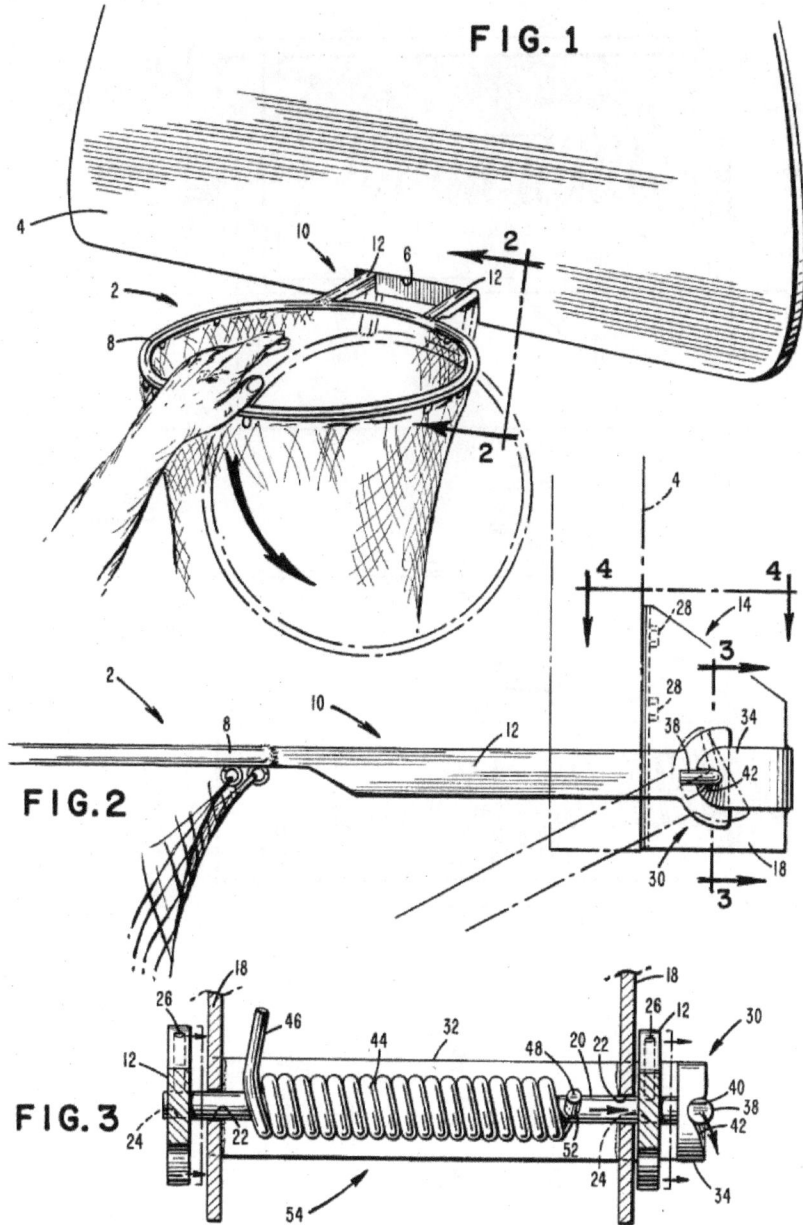

FIG. 1

FIG.2

FIG.3

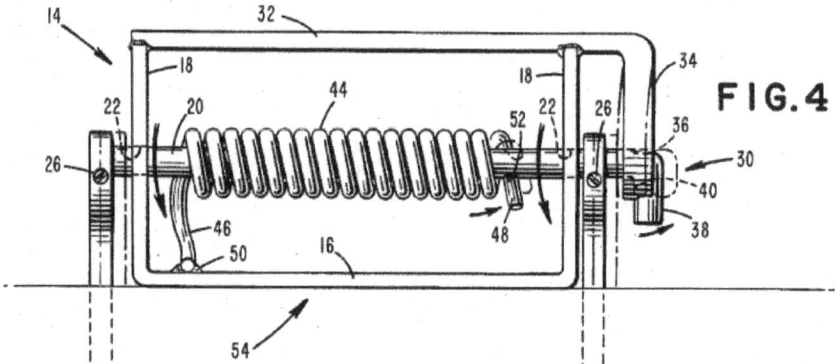

FIG. 4

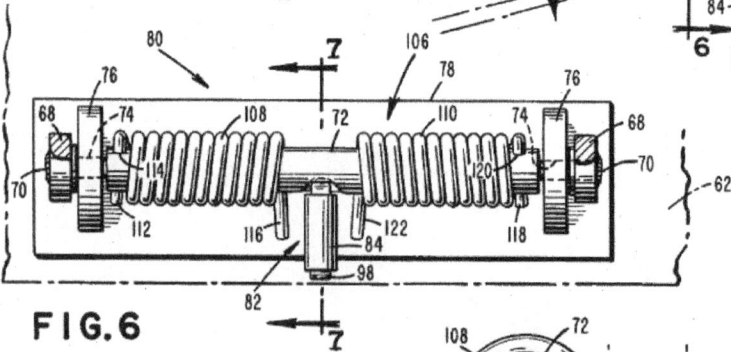

FIG. 5

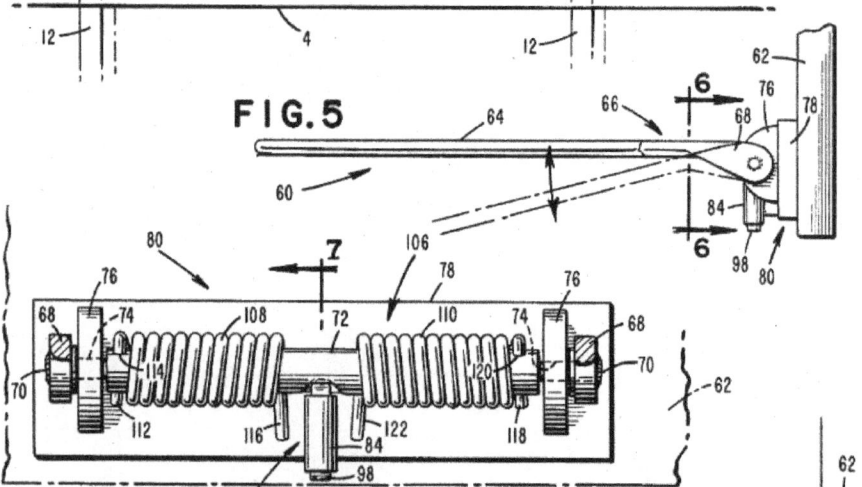

FIG. 6

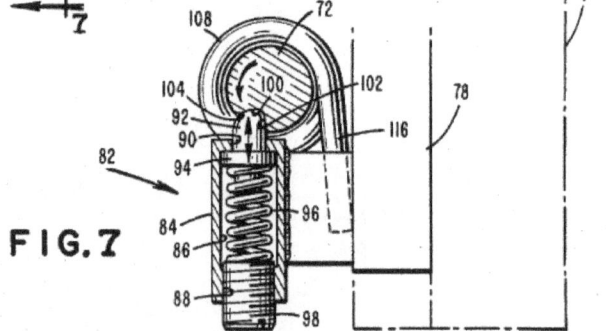

FIG. 7

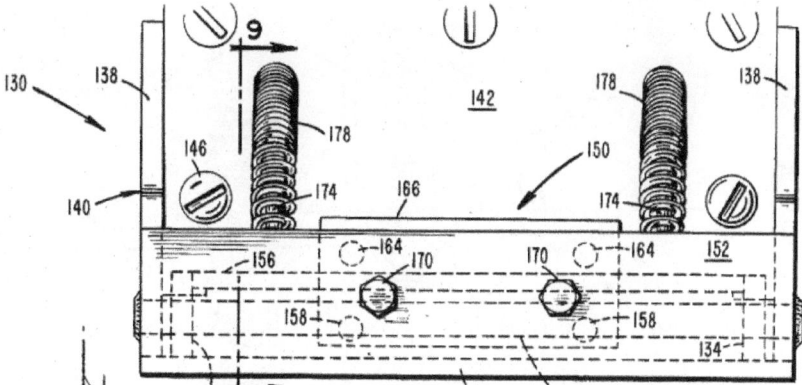

FIG. 8

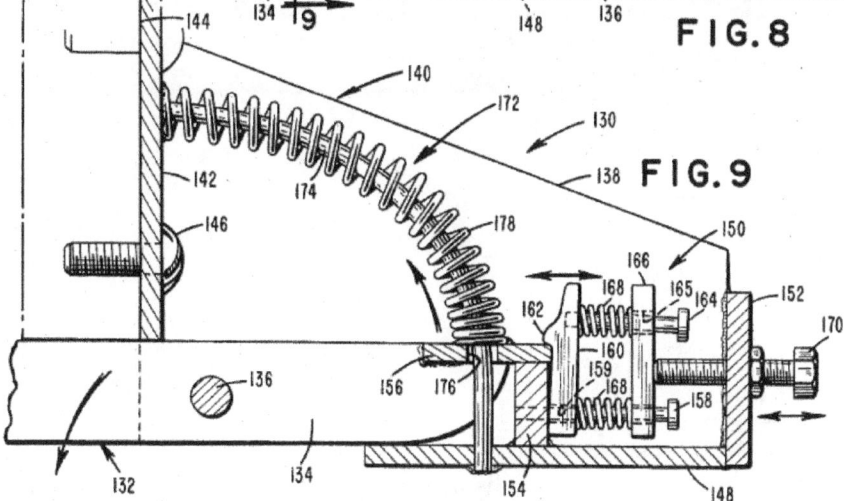

FIG. 9

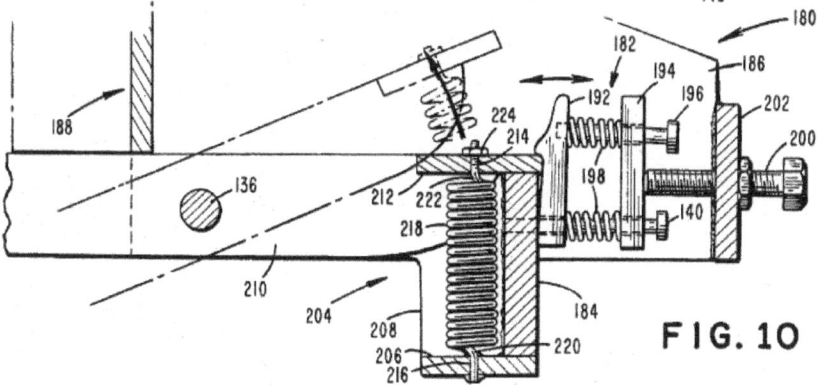

FIG. 10

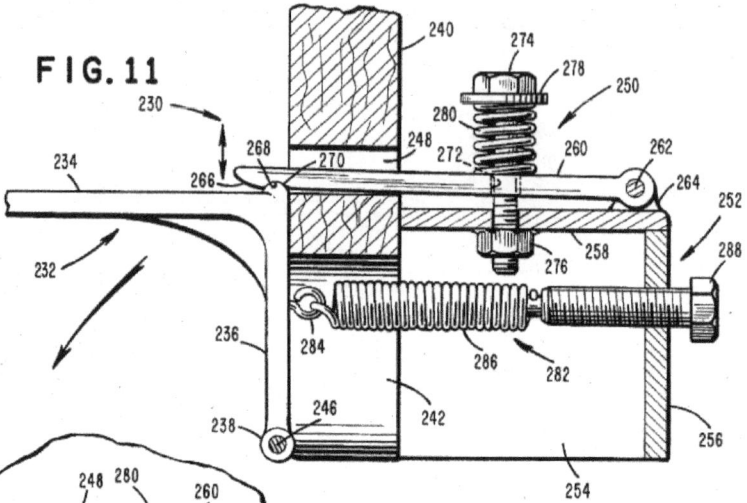

FIG. 11

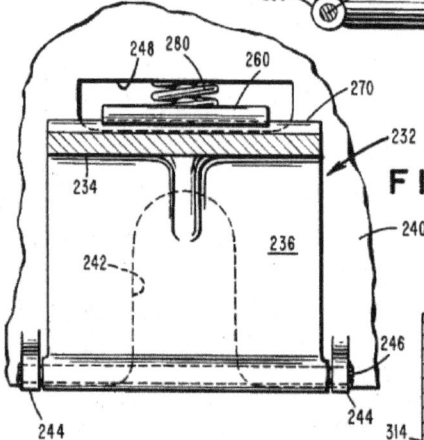

FIG. 12

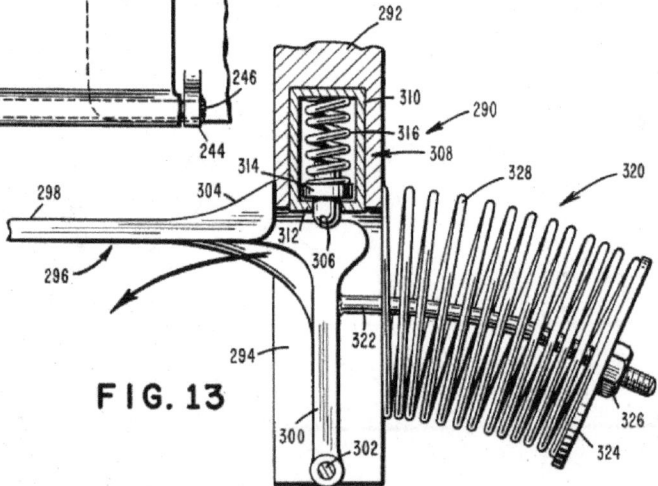

FIG. 13

1

BREAK-AWAY BASKETBALL GOAL

BACKGROUND OF THE INVENTION

1. Field of the Invention

The present invention relates generally to goals for use in playing the game of basketball, and to means for mounting such goals. More particularly, it relates to a new basketball goal arrangement designed so that the hoop thereof can break-away when a predetermined force is applied thereto, and which will return the hoop to its normal horizontal position when the applied force has abated.

2. Description of the Prior Art

The game of basketball is now played world-wide, both indoors and out, by seasoned professionals, high school and college students, and amateurs alike. The popularity of the game is enhanced by the simplicity of its playing equipment, which includes a suitable playing surface on which a basketball court can be established, two goals, one for each end of the court, and a basketball.

Over the years the goal used in the game has emerged from a simple basket, from which the game drew its name, to a circular metal hoop 18 inches in diameter, and which normally carries a mesh net thereon designed to momentarily check the basketball as it passes downwardly through the hoop. The hoop is mounted on or just beneath a backboard, and the backboard is mounted in an elevated position so that the upper edge of the hoop is 10 feet above the playing surface.

While the circular hoop itself is standardized world-wide, a number of different devices have been proposed for mounting thereof. Usually, the hoop is provided with supporting arms that are secured to a mounting bracket, and the latter will normally be secured directly to the face of the backboard. In other instances, the hoop supporting arm can be secured to structure that is itself separate from the backboard, as is shown in U.S. Pat. No. 3,462,143, for example.

Regardless of the mounting structure for the hoop, it is normal practice to mount it so that it is rigidly secured in a horizontal position. The hoop must be sufficiently rigid so that it can withstand the several forces applied thereto during a game, created by the ball bouncing off the rim, or by players coming in contact therewith, and the like.

The rigid basketball hoop has been in use for decades, and until recently has functioned very well. However, in recent years, as basketball players have become taller, the hoops have been subjected to increasing abuse from contact with the players. In some instances this has resulted in damage to the hoop and or the backboard, and somtimes injury to the players.

A special problem has occurred with the glass backboards now in use in many auditoriums and gymnasiums. The preferable way to mount a basketball hoop is on a bracket that is connected directly to the backboard. When this practice has been followed with glass backboards, it has often occured that the backboard wll be shattered during play when one or more players descend heavily on the rigid hoop. In an effort to solve this problem, the hoop has been mounted separately from the backboard, as in the abovenoted U.S. Pat. No. 3,462,143.

The arrangement of the cited patent does lessen the chance of damage to the backboard, but the new relationship of the hoop to the backboard is considered less

2

than fully desireable by some players. Moreover, the problem of damage to the hoop and injury to a player from the still relatively rigid hoop still exists. In addition, the arrangement of the patent is expensive to construct, and hence is not suitable for widespread application on parkland, school and home playing courts.

All of the present concerns about damage and injuries resulting from basketball play with the conventional rigid hoop have been greatly exacerbated recently by the increasingly physical nature of basketball, and particularly by great reliance on the so-called "dunk shot" as a playing tactic. In the dunk shot, the player jumps upwardly with the ball in his outstretched arms and hands, and then drives it downwardly through the loop. Obviously, the chance of player contact with the hoop during such a dunk shot is quite likely, and indeed such often occurs.

Given today's taller, physically larger, and more aggressive players, and the use of the dunk shot, the hoop of the basketball goal is subject to abuse as never before. It is not unknown for a hoop to be badly bent or simply broken off during play, or for a player to be injured as a result of contact with an unyielding hoop. But thus far, no acceptable replacement has been found for the conventional rigid hoop.

Another problem with the rigid basketball hoop occurs because of today's rising tide of vandalism, particularly in urban areas. Many outdoor basketball courts, in particular, have been rendered unserviceable because individuals have jumped up, grabbed on to the hoop, and bent it downwardly. No economical solution to this problem has yet been found, and thus replacement of the damaged goal is usually the only alternative. With public agencies finding increasing difficulty in meeting their budgets, these replacement costs are becoming an increasing burden.

There is a need for a new kind of basketball goal, one that can accommodate today's playing conditions, and in particular the dunk shot, and which is also relatively vandal proof. The present invention is directed toward such a basketball goal.

SUMMARY OF THE INVENTION

The present invention provides a basketball goal that can be utilized for normal play of the game, and wherein the goal is usually rigid, and responds in the normal manner as a basketball strikes it and bounces off. However, should a force in excess of a predetermined value be applied to the hoop, as can occur if the player hangs on the hop during a dunk shot or a vandal deliberately grabs the hoop and hangs on, the hoop will break-away. After the applied force has abated, the hoop will be returned to its normal, horizontal position.

This new concept in a basketball goal practically eliminates damage to the hoop or the backboard during normal play, and greatly reduces the possibility of a player being injured because of collision with the basket during a dunk shot or the like. Further, the invention is of value in reducing damage from vandalism, and indeed eliminates one of the more common causes of vandalism damage.

In the invention, a regulation hoop is provided with a supporting arm structure, and the latter is pivotally mounted so that the hoop can pivot downwardly from a normal, horizontal position, to a break-away position. A releasable detent mechanism is constructed and arranged to normally secure the hoop in its horizontal

3

position, but will release when force in excess of a given value is applied to the hoop, whereupon the hoop will pivot downwardly.

The detent mechanism can assume several different configurations, and in certain embodiments of the invention can be adjusted to accomodate different values for the applied force that will initiate release. This feature makes it possible to set a selected applied force value for each goal that is identical to a prescribed standard, such as would normally be required for tournament or professional play. In all instances, the detent mechanism is effective to hold the hoop rigid during normal play, so that a basketball impinging thereon will bounce off in the usual manner.

In certain use situations, it would be acceptable to provide for manual return of the hoop from its break-away position, to its normally horizontal position. For example, in an inexpensive basketball goal for home or light outdoor use, this might prove acceptable. Under such an arrangement, once the hoop has moved to its break-away position, the user can simply swing it back up into its normal horizontal position, the detent mechanism of the invention being constructed so that it will re-engage during such movement.

For nearly all installations, however, it would be preferable for the hoop to return automatically from its break-away to its normal position. This will avoid interruptions to the game, which could be disruptive and time consuming. Further, in the case of vandalism activities, it would eliminate the need for an attendant to periodically visit the basketball court. The present invention includes a reset mechanism that will accomplish this automatic returning of the hoop to its normal position.

The reset mechanism can obtain its motivating force from a plurality of sources. For example, a hydraulic motor might be employed, or an electrical actuator of the solenoid or some other type could be used. In the latter case, the same solenoid could even function as the detent mechanism, if desired. However, it has been found that the simplest and most economical arrangement is to use a resilient member as a motivating device, and in particular a metallic spring.

The preferred embodiments of the invention utilize tension, compression or torsion spring members, depending upon the particular embodiment. In each case, the spring member is mounted and arranged so that it is effective to return the hoop from its break-away position, and at the same time cause the detent mechanism to be re-engaged.

It is the principle object of the present invention to provide a basketball goal arrangement including a hoop that is mounted to remain in a horizontal position during normal play, but which can break-away when a force in excess of a preselected value is applied thereto.

A further object is to provide a basketball goal arrangement with a break-away hoop, and which includes means to automatically reset the hoop from its break-away to its normal horizontal position.

Another object is to provide a break-away basketball goal that can be economically constructed, and which is relatively maintenance free.

Yet another object is to provide a break-away basketball goal incuding a detent mechanism for releasably securing the hoop of the goal in its normal position, the detent mechanism being designed to provide for normal play.

4

Still another object is to provide a detent mechanism for a break-away basketball goal that can be set to accomodate a selected break-away force.

It is also an object of the invention to provide a reset mechanism for a break-away hoop, designed to return the hoop to its normal position, and at the same time to re-engage the detent mechanism securing the hoop.

Other objects and many of the attendant advantages of the invention will become readily apparent from the following Description of the Preferred Embodiments, when taken in conjunction with the accompanying drawings.

BRIEF DESCRIPTION OF THE DRAWINGS

FIG. 1 is a fragmentary, perspective view showing a backboard equipped with a first embodiment of the break-away goal of the invention, the hoop of the goal being shown in its break-away position by phantom lines;

FIG. 2 is an enlarged, fragmentary side elevational view of the goal assembly of FIG. 1, taken generally along the line 2—2 in FIG. 1, and showing the supporting arm structure in both the normal and break-away positions thereof;

FIG. 3 is a sectional view taken generally along the line 3—3 in FIG. 2, and shows the arrangement of the torsion spring and the shaft of this embodiment;

FIG. 4 is a top plan view of the goal assembly of FIG. 1, taken generally along the line 4—4 in FIg. 2, and showing in particular the mounting bracket and the detent mechanism elements carried thereby;

FIG. 5 is a fragmentary, side elevational view showing a second embodiment of the goal assembly of the invention;

FIG. 6 is an enlarged, fragmentary, front view of the goal assembly of FIG. 5, taken generally along the line 6—6 in FIG. 5, and showing the arrangement of the two torsion springs and the center mounted detent mechanism;

FIG. 7 is an enlarged sectional view taken along the line 7—7 in FIG. 6, and showing the construction of the adjustable detent pin;

FIG. 8 is a rear elevational view of a third embodiment of the break-away goal assembly of the invention, showing in particular the spaced bow springs and the adjustable detent mechanism;

FIG. 9 is a side elevational sectional view of the embodiment of FIG. 8, taken generally along the line 9—9 in FIG. 8, and showing details of the supporting arm structure and how such cooperates with the adjustable detent mechanism;

FIG. 10 is a view similar to FIG. 9, but showing a modification of the embodiment of FIGS. 8 and 9 wherein the two bow springs of the reset mechanism are replaced with a tension spring mounted on a bracket extending downwardly from the supporting arm structure;

FIG. 11 is a fragmentary, side elevational view, partly in section, of a fourth embodiment of the break-away goal assembly of the invention, showing in particular the adjustable tension spring arrangement for biasing the hinged, L-shaped supporting arm structure toward its normal position, and the adjustable detent mechanism;

FIG. 12 is a fragmentary, front elevational view, partly in section, of the goal assembly of FIG. 11; and

FIG. 13 is a view similar to FIG. 11, showing a modification of the fourth embodiment of the invention wherein an L-shaped supporting arm structure is pivoted centrally of a slot opening in a modified backboard, the tension spring reset mechanism of FIG. 11 being replaced by a compression spring mechanism, and the spring-biased detent member being carried by the modified backboard.

DESCRIPTION OF THE PREFERRED EMBODIMENT

As will be appreciated from a review of the drawings, the present invention can be constructed in many different embodiments. The break-away goal can be carried by a bracket that is mounted on the rear of a backboard, or it can be mounted directly on the face of a conventional backboard. Further, in other embodiments of the invention the break-away goal is pivoted directly to a modified backboard. The particular embodiment chosen for use will depend on many factors, including whether a conventional backboard is to be replaced, the type and quality of basketball play that is expected, and the fiscal limitations placed on the user facility. All embodiments of the invention, however, are based on the same inventive concept.

Referring now to FIGS. 1-4, a first embodiment of the break-away basketball goal of the invention is shown generally at 2, mounted on a backboard 4 having a centrally disposed notch 6 in the lower edge thereof. The goal 2 includes a circular hoop 8 having supporting arm structure 10 connected thereto, which structure includes a pair of spaced, generally parallel support arms 12.

The goal 2 is connected to the backboard 4 by a mounting bracket 14, which includes a base plate 16 having a pair of spaced end flanges 18 projecting normally therefrom. A horizontally disposed shaft 20 is rotatably mounted within aligned bores 22 in the end flanges 18, and the opposite ends thereof are received through bores 24 in the outer ends of the support arms 12 and are secured to said arms by set screws 26, or other suitable means. If desired, arms 12 can simply be welded to the shaft 20.

The base plate 16 of the bracket 14 is secured to the rear face of the backboard 4 by bolts 28, whereby the hoop 8 is pivotally mounted for movement between a normally horizontal position, shown by full lines in FIGS. 1 and 2, and a break-away position, shown by phantom lines in said FIGS. The hoop 8 is held in its normal, horizontal position by a detent mechanism, indicated generally at 30.

The detent mechanism 30 includes a bar 32 welded to the end flanges 18, and which extends parallel to the shaft 20. The bar 32 has a resilient arm 34 on one end thereof, which extends parallel to one of the end flanges 18, and which has a bore 36 therein through which one end of the shaft 20 projects, said shaft end having a bent portion 38 thereon that extends normally to the axis of the shaft. The shaft bent portion 38 engages the outer face of the resilient arm 34, and the latter has a groove 40 therein extending radially from the bore 36 and which is shaped to seat the round shaft portion 38.

The groove 40 is so positioned that the shaft bent portion 38 is seated therein when the hoop 8 is in its normal, horizontal position. The downward side of the groove 40 flows into a tapered cam surface 42, on which the bent shaft portion 38 rides when the hoop 8 is moved from its normal, horizontal position toward its break-away position, and return.

A coil spring 44 is wrapped about the shaft 20 between the end flanges 18, and includes first and second end portions 46 and 48. The first end portion 46 engages the base plate 16, and is secured thereto by a weld 50, or other suitable means. The second spring end 48 is passed through a bore 52 provided in the shaft 20. Thus, the coil spring 44 acts like a torsion spring, when the hoop 8 is moved toward its break-away position.

The coil spring 44 is a part of both the detent mechanism 30, and a reset mechanism, indicated generally by the reference number 54. As a part of the detent mechanism 30, it functions to pull the shaft 20 and the supporting arm structure 10 connected thereto in a direction to seat the bent shaft portion 38 in the groove 40, the spring 44 being placed under tension for this purpose. The amount of tension in the spring will directly determine the amount of force that can be applied to the hoop 8, before the detent mechanism 30 releases and the hoop is allowed to move into its break-away position. Thus, by properly choosing the spring 44, the applied break-away force for the goal 2 is determined.

Actually, the resilient arm 34 also plays a roll in the detent mechanism 30, in that it is also deflected during break-away of the hoop 8, as indicated by phantom lines in FIG. 4. This fixing of the arm 34, coupled with axial movement of the shaft 20 against the pull of the spring 44, allows downward pivoting motion of the hoop 8 to occur.

If desired, the device can be constructed so that the resilient arm 34 is the principal resilient force in the detent mechanism 30. To accomplish this, the spacing between the arms 12 is reduced until it is just barely greater than the distance between the outer faces of the end flanges 18. This will eliminate substantially all axial shifting of the shaft 20, and will thereby greatly reduce the roll of the coil spring 44 in the detent mechanism 30. Instead, the resilient arm 34 will then function to retain the hoop 8 in its normal horizontal position, and it will need to be constructed of a spring steel suitable for this function.

A disadvantage of utilizing only the resilient arm 34 to retain the hoop 8 is that precise selection of the applied break-away force required to overcome the detent mechanism is somewhat difficult, being dependent on the material selected for the bar 32 and the precision used in manufacturing the device. However, this arrangement is economical, and rugged in use.

The way in which the detent mechanism functions in FIGS. 1-4 is believed obvious, the mechanism 30 being effective to hold the hoop 8 in its erect position until a sufficient force has been applied to the hoop. Thereupon, the hoop 8 will be swung donwardly, into its break-away position. The reset mechanism 54 is then effective to return the hoop 8 to its original position.

As the shaft 20 rotates during downward movement of the hoop 8, the coil spring 44 is wound in torsion. When the applied deflecting force abates, this torsion spring force is effective to rotate the shaft 20 in the opposite direction, effecting return of the hoop 8. As this return movement accelerates, a substantial force will be generated that will be sufficient to seat the shaft bent position 38 in the groove 40, stopping upward movement of the hoop 8, and fixing the hoop in its normal, horizontal position. It is thus seen that the coil spring 44 functions in tension as a part of the detent mechanism 30, and in torsion as the driving force in the reset mechanism 54.

Turning now to FIGS. 5–7, a second embodiment of the break-away basketball goal of the invention is shown generally at 60, the goal 60 being mounted directly on the front surface of a backboard 62, and including a circular hoop 64 having supporting arm structure 66 attached thereto. The supporting arm structure 66 includes a pair of spaced support arms 68, which are fixed to the reduced diameter opposite end portions 70 of a shaft 72.

The shaft 72 is received through aligned bores 74 provided in spaced flanges 76 carried by a base plate 78, the flanges 76 and the base plate 78 comprising a mounting bracket assembly 80 that is secured to the backboard 62 by suitable means, say by bolts, or even welding, in the instance of steel backboard. The hoop 64 is thus pivotally mounted on the backboard 62, for movement between its normal, horizontal position and a downward, break-away position, as shown in FIG. 5.

The hoop 64 is held in its normal, horizontal position by a detent mechanism 82, which includes a vertical housing 84 mounted centrally on the base plate 78 beneath the shaft 72. The housing 84 has a bore 86 therein, one end 88 of which is threaded, and the other end of which terminates in a reduced diameter opening 90 through which the cylindrical body 92 of a flanged member 94 projects. A coil spring 96 is received within the bore 86 beneath the flanged detent member 94, and is held in place by a screw 98. By adjusting the position of the screw 98, the pressure exerted on the detent member 94 by the coil spring 96 can be adjusted to any selected value, over the rating of the spring.

The shaft 72 has a groove 100 positioned centrally thereof, which includes a radial wall 102 that functions as a stop, and a curved cam wall 104. The head of the flanged detent member 94 is engageable with the groove 100, to releasably secure the hoop 64 in its normal, horizontal position. The detent mechanism 82 will function to hold the hoop 64 horizontal during normal play. When a sufficient deflecting force is applied to the hoop 64 to overcome the compressed coil spring 96, the curved cam wall 104 will depress the flanged detent member 94, whereupon the hoop will pivot downwardly toward its break-away position. By adjusting the position of the screw 98, the hoop can be adjusted to break-away at a selected applied deflecting pressure.

Once the applied force has abated, the hoop 64 can be returned to its normal, horizontal position. This can be done manually, if desired, simply by swinging the hoop 64 upwardly until the detent member 94 is re-engaged in the groove 100, the radial wall 102 assuring accurate positioning of the hoop. However, FIGS. 5–7 include an automatic reset mechanism for returning the hoop 64, such being indicated generally at 106.

The reset mechanism 106 includes a pair of coil springs 108 and 110, both received on the shaft 72 between the flanges 76 on opposite sides of the detent mechanism 82. The coil spring 108 includes a first end portion 112 that is received through a bore 114 in the shaft 72, and a second end portion 116 that engages the base plate 78. Similarly, the second coil spring 110 includes a first end portion 118 engaged in a bore 120 in the shaft 72, and a second end portion 122. The second end portions 122 and 116 can be secured to the base plate 78, if desired, and the coil springs 108 and 110 are arranged so that both will be wound during downward movement of the hoop 64.

As will be evident, the reset mechanism 106 will function automatically to return the hoop 64 to its nor-

mal, horizontal position. In so doing, the momentum built up during upward movement of the hoop 64 will be sufficient at the end of the movement to re-engage the detent mechanism 82, whereupon normal basketball play can resume.

A third embodiment of the invention, utilizing compressed springs as a motivating force in the reset mechanism instead of torsion springs, is shown in FIGS. 8 and 9. The goal of the third embodiment is indicated generally at 130, and includes a hoop (not shown) having supporting arm structure 132 attached thereto, the supporting arm structure 132 including a pair of spaced support arms 134 pivotally mounted on a shaft 136 carried by the spaced end flanges 138 of a mounting bracket 140. The mounting bracket 140 includes a base plate 142 that is secured to the rear surface of a backboard 144 by bolts 146, or other suitable means, and has a bottom plate 148 that extends between the end flanges 138 and which functions as a stop to engage the ends of the support arms 134 to accurately position the hoop in its horizontal position.

A detent mechanism is indicated generally at 150, and includes a front plate 152 carried by the end flanges 138 of the bracket 140, and a mounting plate 154 that is welded to the upper surface of the bottom plate 148 in front of the outer ends of the support arms 134. The support arms 134 are connected by a bridging member 156, which is adapted to rest on the mounting plate 154 when the hoop is in its normal, horizontal position.

The mounting plate 154 carries a pair of headed members 158, which pass through bores 159 provided in a detent bar 160 to mount the bar for sliding movement toward and away from the mounting plate 154. The detect bar 160 carries a detect 162 on its rear face, which is engageable over the bridging member 156. The detent bar 160 also carries a pair of headed members 164, and the headed members 158 and 164 pass through bores 165 provided in a pressure plate 166, whereby the latter is slidably mounted.

Coil springs 168 are mounted on the headed members 158 and 164 and are compressable between the detent bar 160 and the pressure plate 166. A pair of pressure-applying bolts 170 is carried by the front plate 152, the bolts 170 being engageable with the pressure plate 166 to effect compression of the coil springs 168. As will be appreciated, the applied deflecting force required to effect release of the detent mechanism 150 can be easily set, merely by adjusting the positions of the bolts 170. The detent mechanism 150 is at once rugged in construction, and susceptible to being finely set as to release pressure. Thus, it is suited for use where basketball of a high quality is to be played, and precise release pressures are desired for both basketball goals on the court.

The goal 130 includes a reset mechanism 172, having a pair of arched or bowed guide members 174 that are connected at one end thereof to the base plate 142, and at their other end to the bottom plate 148. The guide members 174 pass through bores 176 provided in the bridging member 156, and are shaped so that upward pivoting movement of the outer ends of the support arms 134 is accomodated. Mounted on each guide member 174 is a compression spring 178, and it is believed obvious how the compression springs function to assure return of the hoop to its horizontal position, and re-engaging of the detent mechanism 150.

Referring now to FIG. 10, a modification of the embodiment of FIGS. 8 and 9 is shown, wherein the reset

9

4,534,556

mechanism is modified to employ a tension spring, rather than the compression bow springs 178.

The break-away basketball goal of FIG. 10 is indicated generally at 180, and includes a detent mechanism 182 constructed on the same principles as the detent mechanism 150, except that the mounting plate 184 has a greater height than the mounting plate 154. The mounting plate 184 is welded between the end flanges 186 of a mounting bracket 188, and carries spaced headed members 190 that mount a detent bar 192 and a pressure bar 194. The detent bar 192 carries headed members 196, and the members 190 and 196 carry compression springs 198. Pressure-applying bolts 200 are carried by the front plate 202 of the bracket 188, and function like the bolts 170 to set the release value for the detent mechanism.

The goal 180 includes a reset mechanism 204 that includes a spring support bar 206, welded to the lower edge of the mounting plate 184, and to extensions 208 of the end flanges 186 of the bracket 188. The ends of the support arms 210 carry a bridging member 212 that is engaged by the detent bar 192, and the bridging member 212 and the support bar 206 have aligned bores 214 and 216, respectively, therein. A coil tension spring 218 is disposed between the bar 192 and the member 212, and includes a lower end portion 220 that is passed through the bore 216, and anchored.

The other end of the spring 218 has a threaded end portion 222 thereon, that is received through the bore 214. A nut 224 is received on the end portion 22, and the force of the tension spring 218 can be adjusted merely by adjusting the position of the nut. The manner in which the reset mechanism 204 functions is believed obvious from the drawings.

Turning now to FIGS. 11 and 12, another embodiment of the break-away basketball goal of the invention is shown generally at 230, and includes a supporting arm structure 232 that supports a hoop (not shown), and which is in the form of an L-shaped member having a normally horizontal portion 234 to which the hoop is attached, and a generally vertical base portion 236 that terminates at its lower end in one-half of a hinge 238. A backboard 240 is provided having a vertical notch 242 in the lower edge thereof, and ears 244 are provided on the front face thereof on both sides of the notch. The arm member 232 is pivotally connected to the backboard 240 by a shaft 246 passing through the ears 244 and the hinge half 238.

As shown in the drawing, the backboard 240 has a slot 248 therein, disposed over the notch 242. The goal includes a detent mechanism indicated generally at 250, which mechanism includes a bracket 252 attached to the rear side of the backboard 240 and having end flanges 254, a front plate 256, and a top plate 258. A detent bar 260 extends through the slot 248, and is pivotally connected by a shaft 262 and ears 264 to the top plate 258. The forward end of the detent bar 260 has a rounded underlip 266, leading to a groove 268. The rear end of the horizontal portion 234 of the supporting arm structure 232 carries a detent 270 thereon, which is seatable in the groove 268.

In order to hold the detent 270 seated in the groove 268, the detent bar 260 has a bore 272 therein, through which a bolt 274 passes. The bolt 274 threads into a nut 276 welded to the undersurface of the top plate 258, and carries a washer 278. A coil spring 280 is compressed between the washer 278 and the top surface of the de-

tent bar 260, and provides the resilient force for seating the detent 270.

The manner in which the detent mechanism 250 operates is believed obvious. The bolt 274 makes it possible to adjust the pressure with which the detent bar 260 is urged toward the detent 270, so that the applied deflecting pressure at which break-away will occur can be set to a desired value.

The goal 230 includes a reset mechanism, indicated generally at 282. The rear face of the vertical base portion 236 of the arm structure 232 has an anchor 284 thereon positioned centrally of the notch 242, and to which one end of a coil spring 286 is secured. The other end of the coil spring 286 is connected to an adjusting bolt 288, carried by the front plate 256. The bolt 288 makes it possible to adjust the value of the reset force, and the notch 242 provides for movement of the coil spring 286 during pivotal movements of the supporting arm structure 232.

A variation on the arrangement of FIGS. 11 and 12 is shown in FIG. 13, wherein the modified basketball goal is indicated generally at 290, and includes a backboard 292 having a centrally disposed notch 294 in its lower edge, corresponding to the notch 242. In the goal 290 the supporting arm structure 296 is also L-shaped, and includes a generally horizontal portion 298 that supports the hoop (not shown), and a generally vertical portion 300 that is receivable in the notch 294 and pivoted to the backboard 292 by a shaft 302. The horizontal arm portion 298 has a stop 304 thereon which positively seats the hoop in its horizontal position, and carries a groove 306 arranged to cooperate with a detent mechanism 308 carried in the backboard 292 at the top of the notch 294.

The detect mechanism 308 includes a housing 310 having an inturned retaining lip 312 therein, which is effective to retain a flanged detent member 314 within the housing. A coil spring 316 is disposed within the housing 310 behind the detent member 314, and functions to retain the detent member 314 in engagement with the groove 306 under normal playing conditions. When a deflecting force of sufficient magnitude is applied to the hoop of the goal 290, the detent mechanism 308 will release.

The goal 290 also includes a reset mechanism, indicated generally at 320, and including a guide bar 322 carried by the rear face of the vertical base portion 300 of the arm structure 296, the outer end of the guide bar 322 being threaded and receiving a washer 324 and a nut 326. A compression spring 328 of large diameter is seated between the washer 324 and the edges of the backboard 292 on opposite sides of the notch 242. It is believed that the manner in which the reset mechanism 320 functions is readily apparent.

While a number of different embodiments of the invention have been shown, it is apparent that others can also be devised, according to the concept of the invention. Thus, it is to be understood that further modifications and variations of the invention are possible, within the teachings provided herein and the attached claims.

We claim:
1. A backboard and break-away basketball goal assembly, including:
a backboard;
a hoop having supporting arm structure thereon, said supporting arm structure including a normally vertically disposed mounting plate having upper and lower ends;

11

means for pivotally connecting said normally vertically disposed mounting plate of said supporting arm structure to said backboard, including a generally horizontally disposed pivot pin positioned generally at said lower end of said mounting plate, whereby said hoop can be pivoted downwardly about said pivot pin from a normal, horizontal position, to a break-away position;

detent mechanism means carried by said backboard and said mounting plate for releasably securing said hoop in said normal, horizontal position, said detent mechanism means being constructed to disengage when a deflecting force exceeding a predetermined value is applied to said hoop, whereby said applied deflecting force can then pivot said hoop toward said break-away position, and to then re-engage during return of said hoop from said break-away position to said normal, horizontal position, said detent mechanism means including:

a detent projection mounted on the upper end of said mounting plate;

bracket means mounted on said backboard;

a generally horizontally disposed detent arm, the rear portion of said arm being secured to said bracket means, and said arm extending toward said hoop and including a nose portion which is mechanically engageable with said detent projection to releasably secure said hoop in said normal, horizontal position;

at least one of said detent projection and said nose portion of said detent arm having a tapered surface thereon to facilitate reengagement of said detent mechanism; and

resilient means acting on said nose portion of said detent arm, urging it toward said pivot pin and into engagement with said detent projection, said resilient means being adjustable, whereby it can be set to disengage in response to the application of a selected deflecting force;

reset mechanism means carried by said backboard and operably connected with said supporting arm structure, said reset mechanism means being effective to automatically return said hoop from said break-away position to said normal, horizontal position thereof.

2. An assembly as recited in claim **1**, wherein said reset mechanism means includes:

a tension spring means, one end of said spring means being connected with said bracket means, and the other end thereof being connected with said mounting plate, said backboard being provided with opening means through which said tension spring means extends; and

means between said bracket means and said one, connected end of said spring means for adjusting the tension of said tension spring means, said tension spring means providing the force necessary to return said hoop to its normal, horizontal position, and to re-engage said detent mechanism means.

3. A backboard and break-away basketball goal assembly, including:

a backboard;

a hoop having supporting arm structure thereon, said supporting arm structure including a normally vertically disposed mounting plate having upper and lower ends;

means for pivotally connecting said normally vertically disposed mounting plate of said supporting

12

arm structure to said backboard, including pivot pin means positioned generally at said lower end of said mounting plate, whereby said hoop can be pivoted downwardly about said pivot pin means from a normal, horizontal position, to a break-away position;

detent mechanism means carried by said backboard and said mounting plate for releasably securing said hoop in said normal, horizontal position, said detent mechanism means being constructed to disengage when a deflecting force exceeding a predetermined value is applied to said hoop, whereby said applied deflecting force can then pivot said hoop toward said break-away position, and to then re-engage during return of said hoop from said break-away position to said normal, horizontal position, said detent mechanism means including:

detent projection means mounted on one of said upper end of said mounting plate and said backboard; and

engaging means mounted on the other one of said upper end of said mounting plate and said backboard, said engaging means being positioned and arranged to engage said detent projection means when said hoop is in its normal, horizontal position and said mounting plate is in its generally vertical position;

said engaging means being constructed to include at least one resiliently yieldable portion arranged to mechanically engage with and hold said detent projection means, whereby to enable disengagement of said engaging means from said detent projection means when a deflecting force exceeding a predetermined value is applied to said hoop, and reengagement thereof with said detent projection means when said hoop is returned to its normal, horizontal position, the resilient force exerted by said at least one resiliently yieldable portion being settable mechanically, whereby to adjust said assembly to respond to different applied deflecting forces; and

means carried by said backboard and connected with said mounting plate, constructed and arranged to exert a force urging said mounting plate to pivot toward said backboard, whereby to effect automatic re-engagement of said engaging means with said detent projection means after said deflecting force moves said hoop to its break-away position, and application of said deflecting force to said hoop ceases.

4. A backboard and break-away basketball goal assembly, including:

a backboard;

a hoop having supporting arm structure thereon;

means for pivotally connecting said hoop with said backboard, for pivoting movement about a horizontal axis from a normal, horizontal position, to a break-away position;

detent mechanism means carried by said backboard and said hoop and its supporting arm structure for mechanically releasably securing said hoop in said normal, horizontal position, said detent mechanism means being constructed to mechanically disengage when a deflecting force exceeding a predetermined value is applied to said hoop, and to then mechanically re-engage during return of said hoop from said break-away position,

About the Author

Francis B. "Frank" Francois began his professional career in
the U.S. Patent Office in 1956 as a Patent Examiner after
graduating from Iowa State University with a degree in
engineering. Shortly after moving to Washington, D.C.,
Frank began attending night law school at The George
Washington University. He moved on to become a patent
advisor for the Applied Physics Laboratory at Johns
Hopkins University in 1959, was admitted to the Maryland
bar in 1960, and practiced patent and trademark law with
the firm of Bacon and Thomas from 1962–1980.

Frank entered politics in Prince George's County, MD, in
1962, serving as a Judge on the Orphan's Court Judge, as
County Commissioner and as County Councilman. He was
twice elected President of the National Association of
Regional Councils and in 1979–1980 was President of the
National Association of Counties. In 1980, Frank became
Executive Director of the American Association of State
Highway and Transportation Officials (AASHTO) where he
remained until his retirement in 1999.

Frank was elected to the National Academy of Engineering in
1999, and in 2002 he was recognized with an honorary life
membership in the Institute of Transportation Engineers. In
2003 Iowa State University presented him with the Marston
Medal, its highest award for engineering achievement. In 2004
he was named by the American Road and Transportation
Builders Association (ARTBA) as one of America's top 100
private-sector transportation design and construction
professionals of the 20th century. Other honors and awards
include the George S. Bartlett Award (1997), presented by
AASHTO, ARTBA, and TRB; the Theodore M. Matson
Memorial Award (1993); and TRB's W. N. Carey, Jr.,
Distinguished Service Award (1989).

www.ingramcontent.com/pod-product-compliance
Lightning Source LLC
Chambersburg PA
CBHW032030040426
42448CB00006B/798